MACMILLAN/McGRAW-HILL

Math

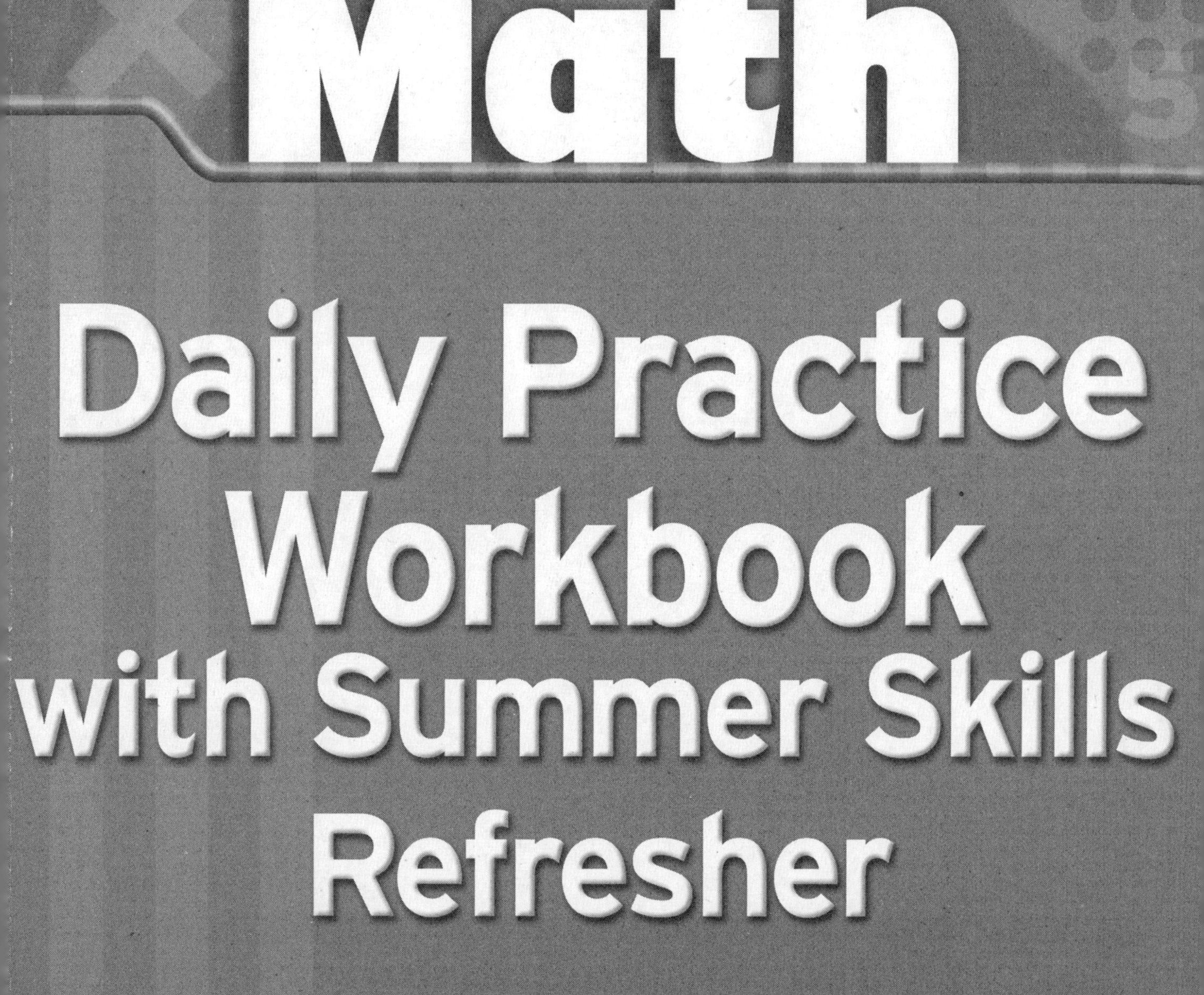

Daily Practice Workbook with Summer Skills Refresher

Grade 3

The McGraw-Hill Companies

Macmillan McGraw-Hill

Published by Macmillan/McGraw-Hill, of McGraw-Hill Education, a division of The McGraw-Hill Companies, Inc., Two Penn Plaza, New York, New York 10121.

Printed in the United States of America

11 12 13 14 HSO 13 12 11 10

Contents

Daily Practice

Summer Skills Refresher

Name ____________________

Counting and Number Patterns

Find the missing numbers. Tell what pattern you used.

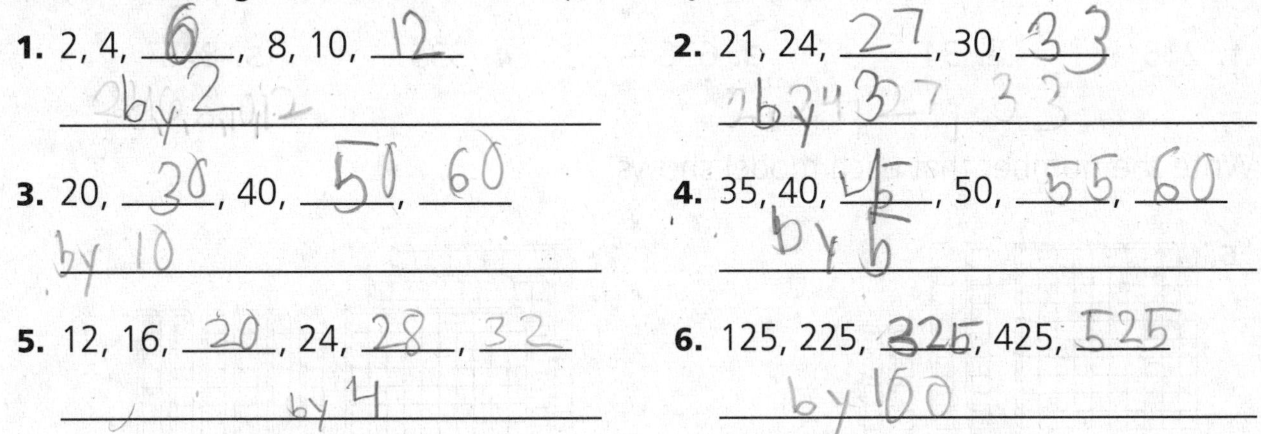

1. 2, 4, ____, 8, 10, ____
   ____

2. 21, 24, ____, 30, ____
   ____

3. 20, ____, 40, ____, ____
   ____

4. 35, 40, ____, 50, ____, ____
   ____

5. 12, 16, ____, 24, ____, ____
   ____

6. 125, 225, ____, 425, ____
   ____

Color the squares that show even numbers yellow.
Color the squares that show odd numbers blue.

7.

1	2	3	4	5	6	7	8	9	10
11	12	13	14	15	16	17	18	19	20
21	22	23	24	25	26	27	28	29	30
31	32	33	34	35	36	37	38	39	40
41	42	43	44	45	46	47	48	49	50

Name ____________________

Explore Place Value

1-2 PRACTICE

Use place-value models to show each number.

1. 215 **2.** 34 **3.** 638 **4.** 333 **5.** 826

Write the number that each model shows.

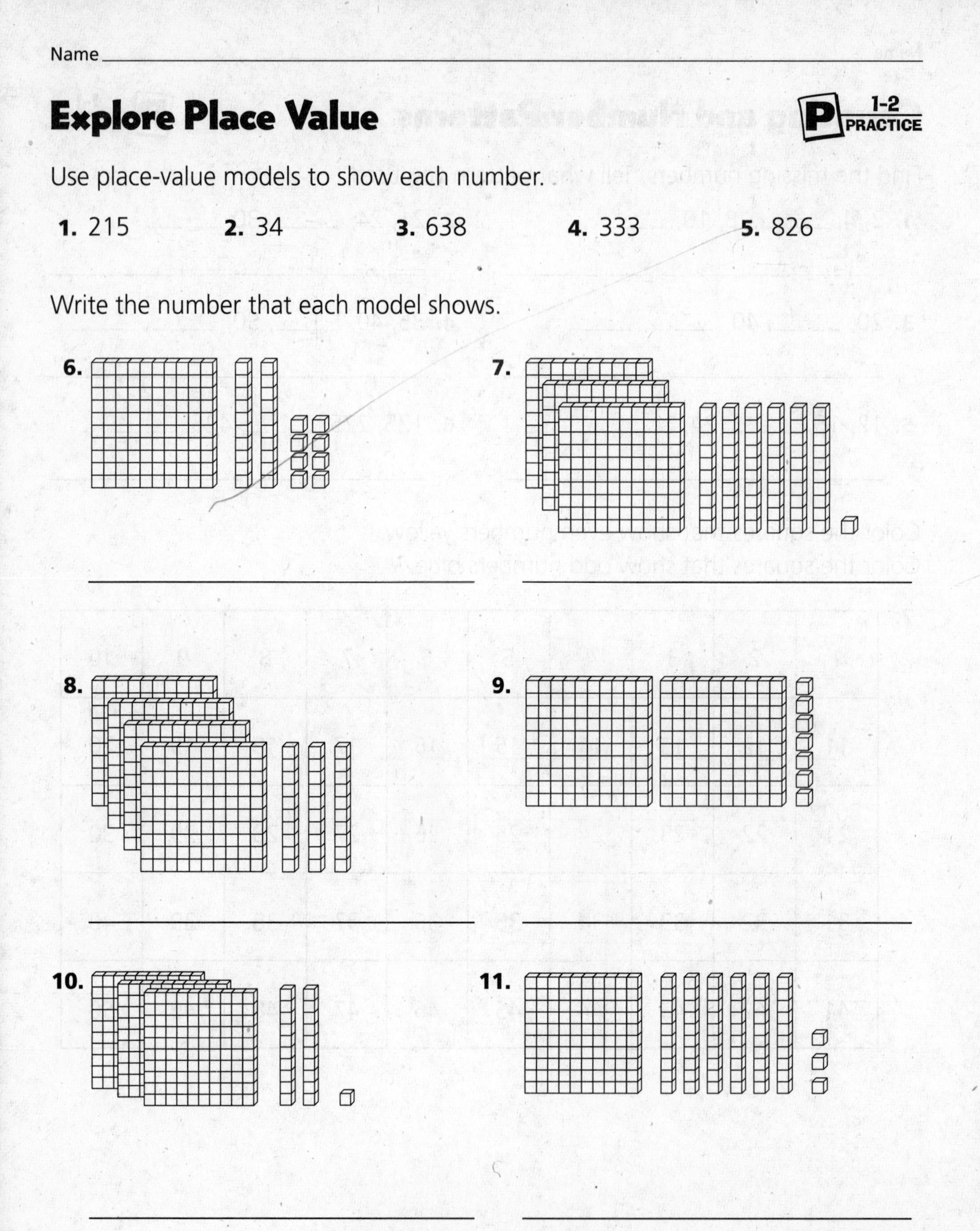

6. ____________________

7. ____________________

8. ____________________

9. ____________________

10. ____________________

11. ____________________

Name ______________________________

Place Value Through Thousands

P 1-3 PRACTICE

Write each number in standard form.

1. 432

2. ______________

3. 600 + 50 + 7 657

4. 5 + 30 + 400 + 2,000 2435

5. six hundred nine 609

6. two thousand, eighty 2080

Write the word form for each number.

7. 374 three hanndred Sevendy-fore

8. 509 ______________

9. 3,800 ______________

10. 6,213 ______________

11. 7,020 ______________

Write each number in expanded form.

12. 293 ______________

13. 407 ______________

14. 4,190 ______________

15. 8,549 ______________

16. 2,050 ______________

Name ______________________________

Place Value Through Hundred Thousands

Write the value of each underlined digit.

1. 4<u>7</u>3 ____________

2. 6,<u>9</u>35 ____________

3. 3,65<u>8</u> ____________

4. 4<u>3</u>,056 ____________

5. 5<u>3</u>6,000 ____________

6. 70<u>5</u>,020 ____________

7. <u>1</u>23,456 ____________

8. 873,<u>4</u>12 ____________

Write the value of 8 in each number.

9. 386 ____________

10. 5,801 ____________

11. 35,618 ____________

12. 298,600 ____________

13. 853,025 ____________

14. 385,060 ____________

Write the digit in each place named.

15. 4,328 (hundreds) _____

16. 62,156 (thousands) _____

17. 503,856 (hundreds) _____

18. 713,604 (hundred thousands) _____

19. 293,418 (ten thousands) _____

20. 375,046 (tens) _____

Algebra Use the rule to complete the table.

21. **Rule** Find the number that is 1,000 less.

Input	Output
4,129	
6,004	
2,995	

22. **Rule** Find the number that is 10,000 more.

Input	Output
38,209	
10,366	
80,004	

Name ______________________________

Explore Money

Count the change you get back from the cashier in a store.
Write the amount and list the coins and bills.

1. Cost: $0.68 You give: $1.00

Your change: ______________________________

2. Cost: $3.45 You give: $5.00

Your change: ______________________________

Complete the table. Use play money to help.

	Toy	Cost	You Give	Change
3.	ball	$0.79	$1.00	
4.	yo-yo	$1.65	$5.00	
5.	jacks	$0.97	$5.00	
6.	truck	$2.83	$5.00	

Name ______________________________

Count Money and Make Change

P 1-6 PRACTICE

Write the money amount.

1. ________

2. ________

3. ________

4. ________

Find the amount of change. List the coins and bills you use.

5. Cost: $1.38
You give: $2.00 Change: ________

6. Cost: $2.08
You give: $5.00 Change: ________

7. Cost: $3.74
You give: $10.00 Change: ________

8. Cost: $7.88
You give: $10.00 Change: ________

Problem Solving

Solve.

9. Ross buys a game for $6.75. He gives the cashier $10.00. How much change does he get?

____

10. Tawana buys a puzzle for $3.45. The cashier gives her $1.55 in change. How much did she give the cashier?

____

Use with Grade 3, Chapter 1, Lesson 6, pages 14–16.

Name ___

Problem Solving: Skill
Using the Four-Step Process

1-7 PRACTICE

Solve.

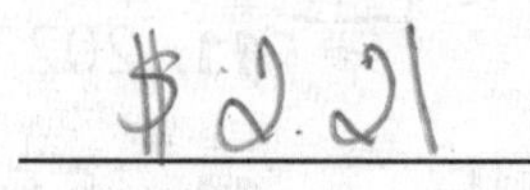

1. Stan hits a target worth 50 points. He then hits a target worth 5 points three times. How many points does Stan have now?

   65

2. Jodie buys a game for $7.79. She gives the clerk two $5-bills. What is her change?

   $2.21

3. Yoko has 400 points. Dan has 200 points less than Yoko. Juan has 300 points more than Dan. Who is the winner?

   Juan

4. Nick starts with $500 in play money. In three rounds of a game, Nick wins $20 in each round. How much play money does Nick have after those three rounds?

   560

5. Susan's game piece is on box 20 of a game board. She moves it ahead 2 boxes, four times. Where is her game piece now?

   28

6. Yolanda has 220 points. Karen has 180 points. Vern has 190 points. Who is the winner?

   220

7. Rick has 4,400 points. He has one turn left. The record is 5,100 points. If Rick scores 500 more points, how many points will he have? Will he break the record?

   no he will not brack the record 4,900

8. Nancy scores 450 points in the first round, 100 points in the second round, and 300 points in the third round. Does she score more than 1,000 points?

   no 850 ponts

Name ________________________________

Compare Numbers and Money

2-1 PRACTICE

Compare. Write >, <, or =.

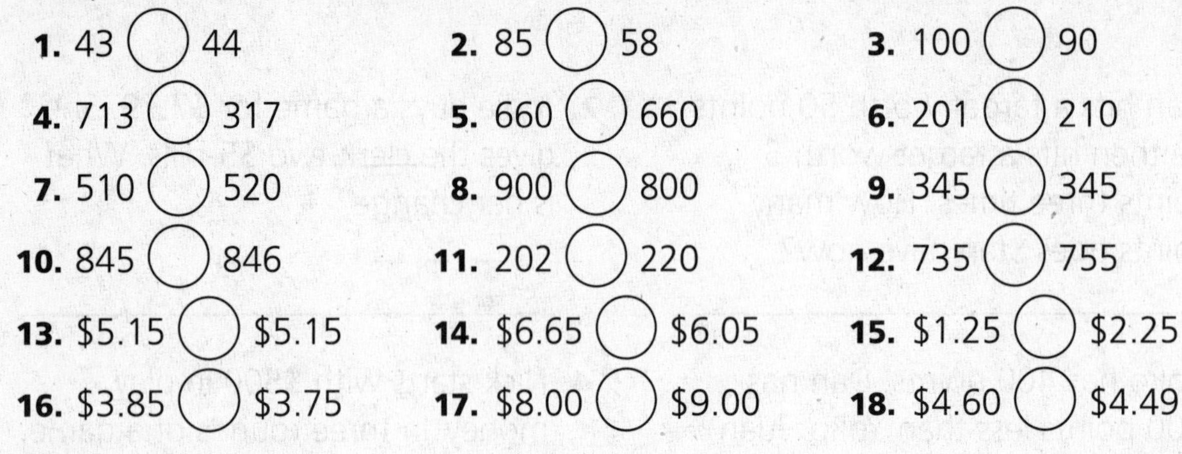

1. 43 ◯ 44 **2.** 85 ◯ 58 **3.** 100 ◯ 90

4. 713 ◯ 317 **5.** 660 ◯ 660 **6.** 201 ◯ 210

7. 510 ◯ 520 **8.** 900 ◯ 800 **9.** 345 ◯ 345

10. 845 ◯ 846 **11.** 202 ◯ 220 **12.** 735 ◯ 755

13. \$5.15 ◯ \$5.15 **14.** \$6.65 ◯ \$6.05 **15.** \$1.25 ◯ \$2.25

16. \$3.85 ◯ \$3.75 **17.** \$8.00 ◯ \$9.00 **18.** \$4.60 ◯ \$4.49

Problem Solving.

Solve.

19. Grace spent \$4.50 on lunch. Amy spent \$5.50. Who spent more money on lunch?

____

20. Alex has 250 coins in his collection. James has 300 coins. Who has fewer coins?

____

21. Kelly has saved \$5.00. Bill has saved \$4.25. Does Bill need more or less money to have the same amount as Kelly? ________

How much more or less? ________

22. Maria took 36 pictures on her vacation. Ellie took 10 more pictures than Maria. How many pictures did Ellie take?

____

Name ____________________

Order Numbers and Money

P 2-2 PRACTICE

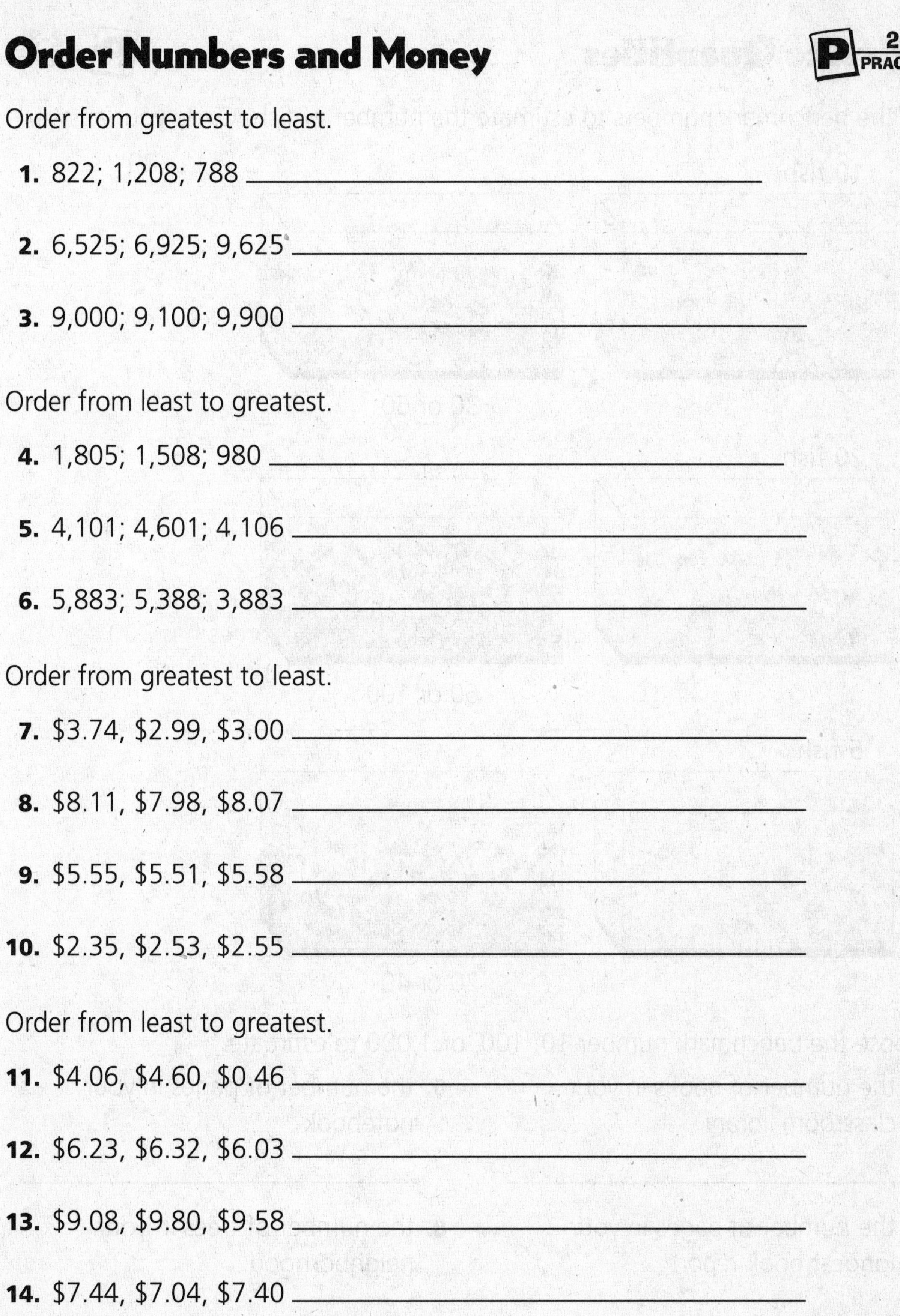

Order from greatest to least.

1. 822; 1,208; 788 ____________________

2. 6,525; 6,925; 9,625 ____________________

3. 9,000; 9,100; 9,900 ____________________

Order from least to greatest.

4. 1,805; 1,508; 980 ____________________

5. 4,101; 4,601; 4,106 ____________________

6. 5,883; 5,388; 3,883 ____________________

Order from greatest to least.

7. $3.74, $2.99, $3.00 ____________________

8. $8.11, $7.98, $8.07 ____________________

9. $5.55, $5.51, $5.58 ____________________

10. $2.35, $2.53, $2.55 ____________________

Order from least to greatest.

11. $4.06, $4.60, $0.46 ____________________

12. $6.23, $6.32, $6.03 ____________________

13. $9.08, $9.80, $9.58 ____________________

14. $7.44, $7.04, $7.40 ____________________

Name ____________________

Estimate Quantities

Use the benchmark numbers to estimate the number of fish. Circle your answer.

1. 10 fish

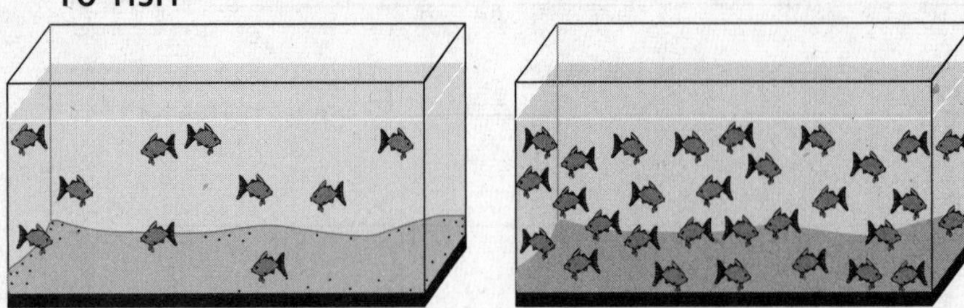

30 or 60

2. 20 fish

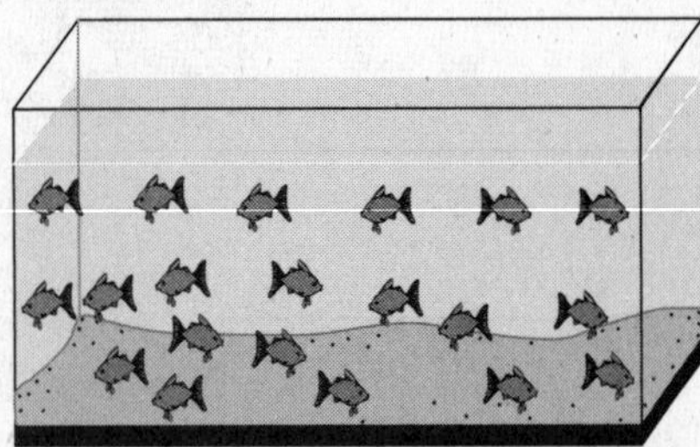

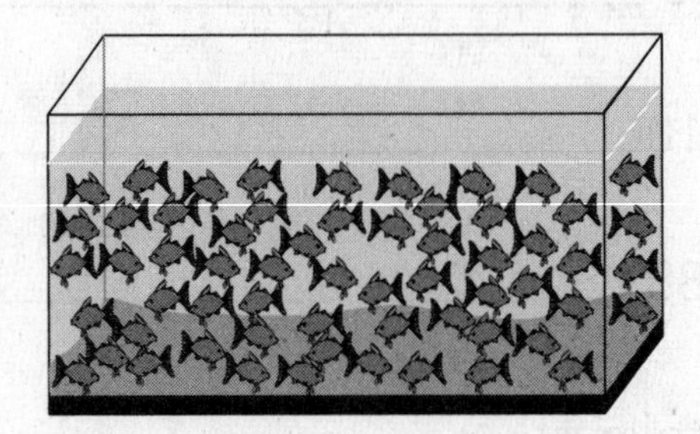

60 or 100

3. 5 fish

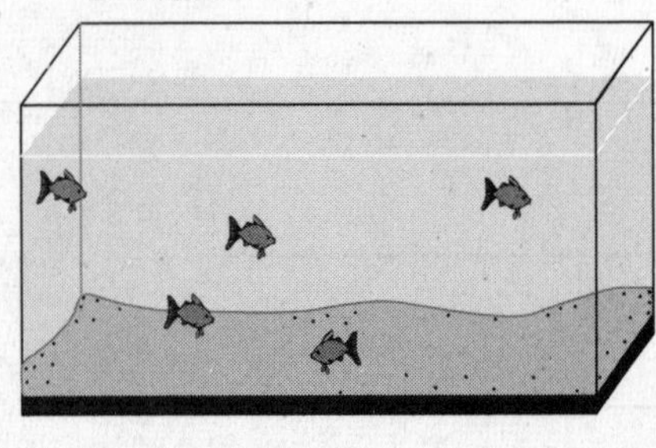

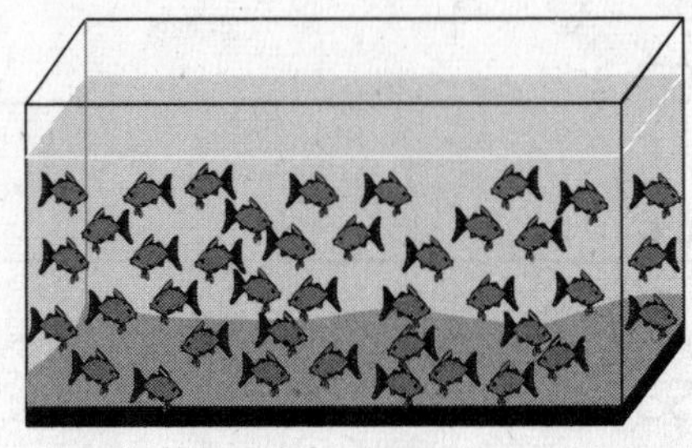

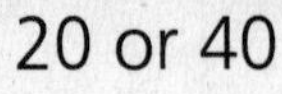

20 or 40

Choose the benchmark number 10, 100, or 1,000 to estimate.

5. the number of books in your classroom library

____

6. the number of pages in your notebook

____

7. the number of pages in your longest book report

____

8. the number of trees in your neighborhood

____

Name ____________________

Round to Tens and Hundreds

P 2-4 PRACTICE

Round to the nearest ten or ten dollars.

1. 37 47

2. $65 70

3. 80 ______

4. 173 170

5. 462 460

6. $358 ______

7. 735 740

8. 2,079 2080

9. 1,398 ______

Round to the nearest hundred or hundred dollars.

10. 217 ______

11. 805 ______

12. $319 ______

13. $873 ______

14. $2,350 ______

15. 9,555 ______

16. 3,508 ______

17. $56,071 ______

18. 27,955 ______

Algebra Find the missing digit to make the sentence true.

19. 6☐8 rounds to 630. ______

20. ☐,524 rounds to 8,000. ______

21. 2☐8 rounds to 200. ______

22. 4,5☐3 rounds to 4,600. ______

Problem Solving

Solve.

23. The video game *TechMaster* costs $38.75. How much is that rounded to the nearest ten dollars?

______

24. The video game *DataGenius* is on sale for $24.95. How much is that rounded to the nearest ten dollars?

______

25. Ms. Jones, the computer teacher, bought 87 video games for her school. To the nearest ten, how many games did she buy?

______

26. Jason scored 213 points playing a video game. To the nearest hundred, how many points did he score?

______

Name ____________________

Round to the Nearest Thousand

Round to the nearest thousand.

1. 1,362 ________ **2.** 9,420 ________ **3.** 2,564 ________

4. 8,410 ________ **5.** 6,742 ________ **6.** $3,500 ________

7. 24,391 ________ **8.** 74,687 ________ **9.** $15,894 ________

Use data from the table for problems 10–13.

Depths of Oceans	
Ocean	**Depth**
Pacific	12,925 ft
Atlantic	11,730 ft
Indian	12,598 ft
Arctic	3,407 ft

10. What is the depth of the Arctic Ocean rounded to the nearest thousand?

____

11. Which ocean has an average depth of about 12,000 ft?

____

12. What is the depth of the Pacific Ocean rounded to the nearest thousand?

____

13. What is the depth of the Indian Ocean rounded to the nearest thousand?

____

Name ____________________

Problem Solving: Strategy

Make a Table

Organize the data below in a table.

My Favorite Game

Computer: Jessica, Michael, Akako, Taylor, Aretha, Jamal, Rick, Paula

Board: Erica, Lauren, Mark, Andrew, Allison

Card: Justin, Carl, Dixie, Ben

Game	Tally	Number
Computer	~~||||~~ |||	8
Board	~~||||~~	5
Card	||||	4

Use the table to solve problems 1 and 2.

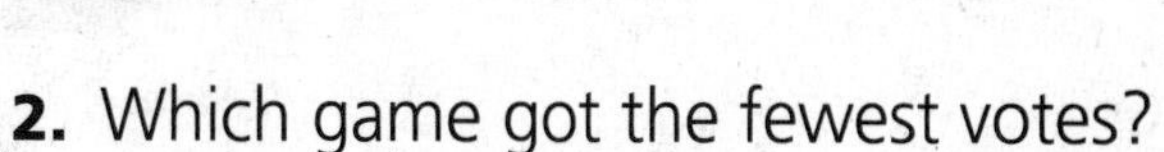

1. Which game got the most votes?

Computer

2. Which game got the fewest votes?

Card

Complete. Lorna draws some shapes: ○ ☆ ○ □ ☆ □ ☆ ☆ ○ □

3. How many more stars than circles did Lorna draw? Make a table in the box.

____

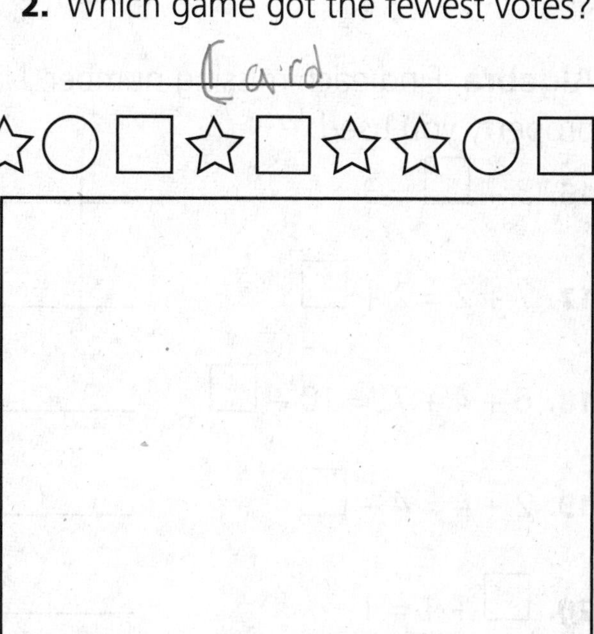

4. Suppose that Lorna draws 2 more squares. How many squares will she have then?

____

Mixed Strategy Review

Solve. Use any strategy.

5. On a dartboard, a wedge has a value of 5 points. Jason hits the wedge 3 times. How many points does he score?

____

Strategy: ____________________

6. **Write a problem** for which you would make a table to solve. Share it with others.

____

____

____

Name ______________________________

Addition Properties • Algebra

Add. Show how you used strategies and properties.

1. $8 + 3 =$ _____ **2.** $9 + 0 =$ _____ **3.** $8 + 9 =$ _____

4. $1 + 0 + 9 =$ _____ **5.** $7 + 3 + 2 =$ _____ **6.** $4 + 2 + 4 =$ _____

7. $3 + 7 + 3 + 1 =$ _____ **8.** $2 + 0 + 8 + 2 =$ _____ **9.** $6 + 2 + 2 + 4 =$ _____

10. $8 + 0$ **11.** $21 + 9$ **12.** $7 + 8$ **13.** $4 + 4$ **14.** $6 + 7$ **15.** $3 + 17$

Algebra Find each missing number. Name the addition property you used.

16. $3 + \square = 3$ ______________________

17. $7 + 2 = 2 + \square$ ______________________

18. $6 + 4 + 7 = 10 + \square$ ______________________

19. $2 + 4 = 4 + \square$ ______________________

20. $\square + 1 = 1$ ______________________

Problem Solving

Solve.

21. Peter buys 15 postcards. The next day, he buys 3 more postcards. How many postcards does he have in all?

____

22. Mr. Tandy has 2 blue shirts, 1 yellow shirt, and 8 white shirts. How many shirts does he have in all?

____

Name ________________________________

Addition Patterns • Algebra

Write the number that makes each sentence true.

1. 3 + 5 = ________

30 + 50 = ________

300 + 500 = ________

2. 6 + 8 = ________

60 + 80 = ________

600 + 800 = ________

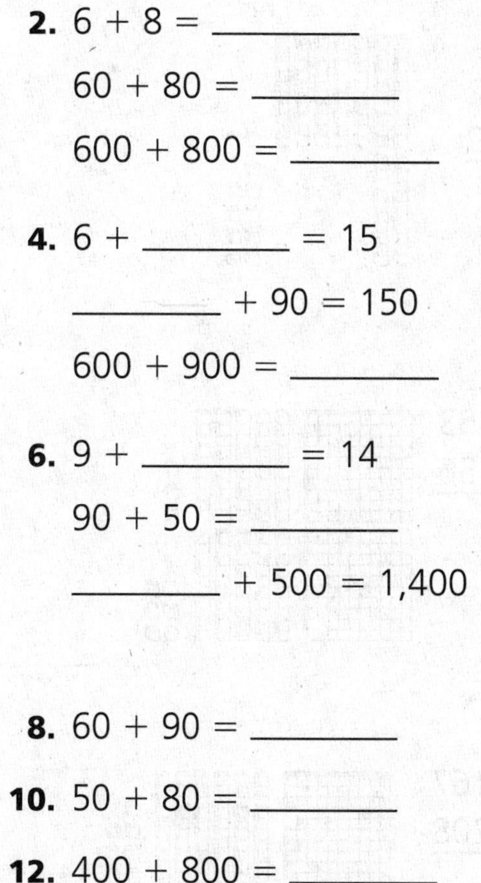

3. 9 + 8 = ________

90 + 80 = ________

900 + 800 = ________

4. 6 + ________ = 15

________ + 90 = 150

600 + 900 = ________

5. ________ + 6 = 13

70 + ________ = 130

700 + ________ = 1,300

6. 9 + ________ = 14

90 + 50 = ________

________ + 500 = 1,400

Add. Use mental math.

7. 40 + 50 = ________

8. 60 + 90 = ________

9. 70 + 80 = ________

10. 50 + 80 = ________

11. 700 + 300 = ________

12. 400 + 800 = ________

13. 900 + 900 = ________

14. 700 + 900 = ________

15. 500 + 90 = ________

16. 800 + 70 = ________

17. 600 + 40 + 80 = ________

18. 600 + 300 + 90 = ________

19. 2,000 + 7,000 = ________

20. 6,000 + 5,000 = ________

Problem Solving

Solve.

21. There are 30 goldfish in one tank and 50 goldfish in another tank. How many goldfish are there altogether?

____

22. The movie theater has 600 seats downstairs and 200 seats in the balcony. How many seats are there altogether?

____

Name ______________________________

Explore Regrouping in Addition

Find each sum.

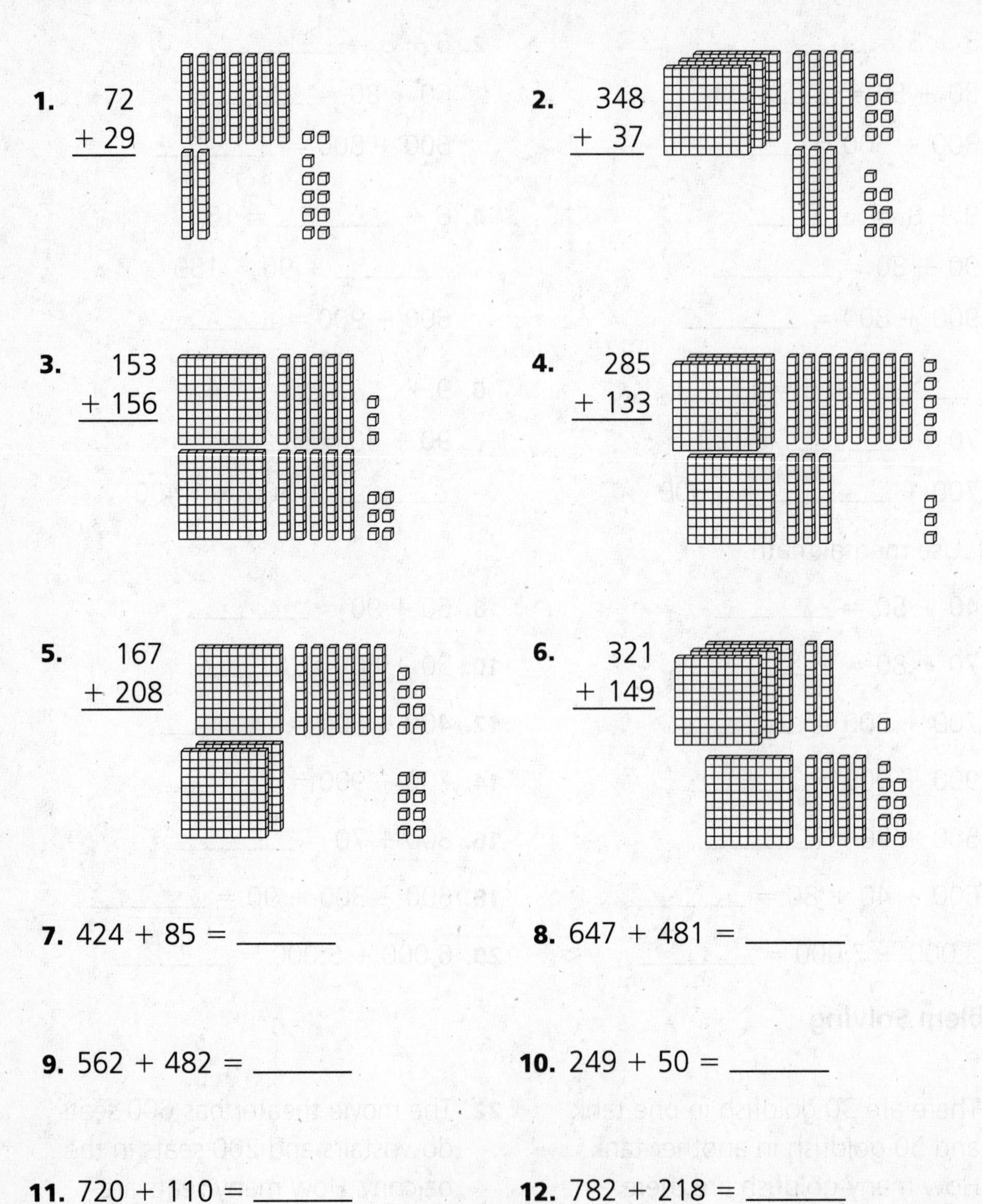

1. 72 + 29 = ______

2. 348 + 37 = ______

3. 153 + 156 = ______

4. 285 + 133 = ______

5. 167 + 208 = ______

6. 321 + 149 = ______

7. 424 + 85 = ______

8. 647 + 481 = ______

9. 562 + 482 = ______

10. 249 + 50 = ______

11. 720 + 110 = ______

12. 782 + 218 = ______

Name ____________________

Add Whole Numbers

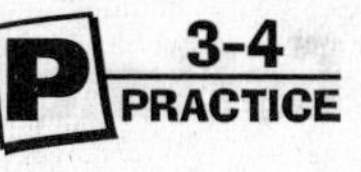

Find each sum.

1. $\begin{array}{r} 233 \\ +\ 556 \\ \hline \end{array}$

2. $\begin{array}{r} 748 \\ +\ 16 \\ \hline \end{array}$

3. $\begin{array}{r} 356 \\ +\ 27 \\ \hline \end{array}$

4. $\begin{array}{r} 623 \\ +\ 48 \\ \hline \end{array}$

5. $\begin{array}{r} 829 \\ +\ 152 \\ \hline \end{array}$

6. $\begin{array}{r} 706 \\ +\ 165 \\ \hline \end{array}$

7. $\begin{array}{r} \$5.00 \\ +\ 1.97 \\ \hline \end{array}$

8. $\begin{array}{r} 477 \\ +\ 304 \\ \hline \end{array}$

9. $\begin{array}{r} 648 \\ +\ 563 \\ \hline \end{array}$

10. $\begin{array}{r} \$964 \\ +\ 39 \\ \hline \end{array}$

11. $\begin{array}{r} 555 \\ +\ 555 \\ \hline \end{array}$

12. $\begin{array}{r} 693 \\ +\ 818 \\ \hline \end{array}$

13. $\begin{array}{r} \$5.62 \\ +\ 2.49 \\ \hline \end{array}$

14. $\begin{array}{r} 276 \\ +\ 576 \\ \hline \end{array}$

15. $\begin{array}{r} 609 \\ +\ 545 \\ \hline \end{array}$

16. 187 + 108 = ________

17. \$1.75 + \$1.20 = ________

18. 297 + 48 = ________

19. 715 + 560 = ________

20. 749 + 857 = ________

21. 550 + 230 = ________

22. 247 + 125 = ________

23. \$6.73 + \$4.52 = ________

Algebra. Find each missing digit.

24. $\begin{array}{r} 67\square \\ +\ 585 \\ \hline 1{,}255 \end{array}$

25. $\begin{array}{r} 3\square5 \\ +\ 29 \\ \hline 394 \end{array}$

26. $\begin{array}{r} 174 \\ +\ 71\square \\ \hline 890 \end{array}$

27. $\begin{array}{r} 578 \\ +\ \square27 \\ \hline 1{,}505 \end{array}$

Problem Solving

Solve.

28. Roberto works at the zoo. He earned \$410 in June and \$375 in July. How much did he earn in all?

____

29. Cathy's class took 150 pictures of crocodiles at the zoo. Joanne's class took 95 pictures of crocodiles. How many pictures were taken in all?

____

Name ____________________

Estimate Sums

3-5 PRACTICE

Estimate each sum. Show how you rounded.

1. 58 + 21

_______ + _______ = _______

2. 64 + 25

_______ + _______ = _______

3. 519 + 388

_______ + _______ = _______

4. 1,749 + 3,902

_______ + _______ = _______

Estimate each sum.

5. 19 + 48 _______

6. 373 + 685 _______

7. 403 + 372 _______

8. 2,099 + 6,785 _______

9. 481 + 88 _______

10. 4,276 + 950 _______

Algebra Write > or < to make the sentences true.

11. 54 + 32 ◯ 80

12. 36 + 59 ◯ 90

13. 45 + 33 ◯ 80

14. 293 + 555 ◯ 800

15. 470 + 381 ◯ 900

16. 251 + 727 ◯ 1,000

17. 329 + 193 ◯ 600

18. 400 + 390 ◯ 700

19. 351 + 490 ◯ 700

20. 702 + 510 ◯ 1,100

21. 382 + 110 ◯ 600

22. 631 + 711 ◯ 1,300

Problem Solving

Solve.

23. The animal shelter has 27 dogs and 49 cats. About how many dogs and cats at the animal shelter?

____

24. There were 581 visitors at the zoo on Saturday and 633 visitors on Sunday. About how many people visited the zoo on the weekend?

____

Name ____________________

Problem Solving: Skill
Estimate or Exact Answer

Choose the best answer.

1. Birdland Park has 525 visitors on Monday. On Tuesday Birdland has 640 visitors. Is the total number of visitors for the two days greater than 1,000?

What do you have to do to solve this problem?

Find the exact sum of 525 + 640.

Estimate to tell if 525 + 640 is greater than 1,000.

Estimate 1,000 + 640.

Solve. Tell if you gave an estimate or the exact answer.

2. Sandy counts 39 birds in the morning and 48 birds in the afternoon. About how many birds does she count altogether?

____

3. The animal shelter has 42 cats, 59 dogs, and 12 birds. Does the shelter have more than 100 animals?

____

4. A pet store has 22 poodles and 35 terriers. There is enough dog food to feed 50 dogs. How many dogs does the store have to feed?

____

5. At the aquarium, Tori counts 16 seals, 14 porpoises, and 21 dolphins. About how many seals, porpoises, and dolphins are there altogether?

____

6. In one year, Central Animal Shelter takes in 264 dogs and 172 cats. Does the shelter take in more than 400 animals?

____

7. Stan's Dude Ranch has 32 horses and 12 ponies. Stan has 52 people coming to the ranch today. How many horses and ponies does Stan have altogether? Does Stan have one horse for each person?

____

Name ____

Add Greater Numbers

4-1 PRACTICE

Add. Check. Is each answer reasonable?

1. 3,186 + 1,812 = ______

2. 6,915 + 2,436 = ______

3. 7,058 + 966 = ______

4. $37.75 + 26.98 = ______

5. 7,304 + 6,783 = ______

6. 7,251 + 748 = ______

7. $56.09 + 9.58 = ______

8. 2,456 + 735 = ______

9. 7,005 + 256 = ______

10. 9,751 + 5,739 = ______

11. $24.36 + $38.45 = ________

12. 4,621 + 5,448 = ________

13. 1,716 + 7,665 = ________

14. $3,390 + $2,285 = ________

15. 9,455 + 2,017 = ________

16. 1,332 + 5,599 = ________

17. 8,724 + 6,114 = ________

18. $35.20 + $32.49 = ________

19. $4,328 + $5,122 = ________

20. 5,026 + 4,563 = ________

Algebra Complete. Write > or <.

21. 4,320 + 5,145 ◯ 3,921 + 5,218

22. 3,962 + 4,331 ◯ 4,120 + 5,096

23. 6,547 + 1,915 ◯ 5,203 + 2,876

24. 7,248 + 3,101 ◯ 7,408 + 2,222

Problem Solving

Solve.

25. At the feed store, Maya Carter spends $86.93 on chicken feed and $23.15 on straw for the chicken house. How much does Maya spend in all?

____

26. On Tuesday, a farmer collected 1,509 eggs from her chickens. The next day she collected 1,531 eggs. How many eggs did the farmer collect altogether?

____

Name ____________________

Add More Than Two Numbers

Add. Check. Is each answer reasonable?

1. $378 + 133 + 69

2. 623 + 95 + 188

3. 123 + 456 + 789

4. $21.26 + 67.45 + 30.23

5. 1,503 + 5,390 + 3,408

6. $237 + 310 + 406

7. 485 + 32 + 310

8. 162 + 263 + 364

9. $207 + 108 + 89

10. 453 + 354 + 534

11. $1,583 + 381 + 626

12. 117 + 224 + 185

13. $9.26 + 4.01 + 5.55

14. 335 + 228 + 445

15. 3,055 + 2,118 + 533

16. 378 + 87 + 300 = ________

17. 255 + 261 + 250 = ________

18. $497 + $852 + $23 = ________

19. 850 + 905 + 109 = ________

20. $62.05 + $57.78 + $82.46 = ________

21. 3,399 + 412 + 74 + 561 = ________

22. 626 + 45 + 890 + 6,572 = ________

23. 7,213 + 181 + 312 + 5,097 = ________

Problem Solving

Solve.

24. Myra buys a fish tank and some equipment. She spends $59 for the tank, $125 for the air pump, and $38 for lighting. How much money does Myra spend?

____

25. Kevin buys a collar for $7.59 for his Dalmatian. He also buys a chew toy for $3.85 and a water bowl for $6.27. How much money does Kevin spend?

____

Name ___

Problem Solving: Strategy

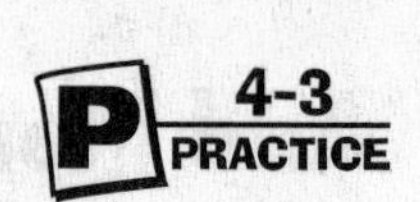

Draw a Diagram

Use the Venn diagram to solve problems 1–3.

1. How many children play soccer only?

___

2. How many children play soccer and basketball?

___

3. How many children play basketball only?

___

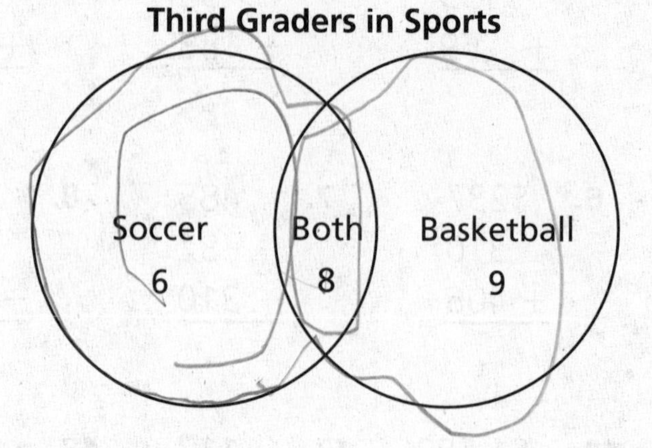

Mixed Strategy Review

Solve. Use any strategy.

4. The basketball team scored 8 points in the first quarter, 9 points in the second quarter, and 5 points in the third quarter. How many points did they score during the three quarters?

___

Strategy: ___

5. The girls' basketball game drew 172 people. The boys' basketball game drew about 163 people. About how many people came to the games altogether?

___

Strategy: ___

6. Show one way to organize the results of the survey.

7. What color was the favorite?

___

Survey What is your favorite color?	
Emma	purple
Sam	red
Nick	green
Ben	purple
Andy	purple
Arthur	red
Paul	purple

Name ____________________

Choose a Computation Method

Add. Tell which method you use.

1. 2,423 + 1,684 = ________

2. 4,523 + 2,397 = ________

3. 8,001 + 1,882 = ________

4. 6,491 + 3,423 = ________

5. 1,896 + 2,812 = ________

6. 5,403 + 4,492 = ________

7. 3,681 + 308 = ________

8. 3,238 + 4,790 = ________

9. 4,084 + 3,234 = ________

10. 781 + 103 + 100 = ________

11. 1,879 + 300 + 3,886 = ________

12. 352 + 719 + 491 = ________

13. 5,923 + 1,336 + 492 = ________

Algebra Find the missing number.

14. 674 + ☐ = 922 ________

15. ☐ + 2,893 = 4,665 ________

16. 982 + ☐ = 2,684 ________

17. ☐ + 443 = 642 ________

18. 506 + ☐ = 900 ________

19. ☐ + 3,606 = 4,887 ________

Problem Solving

Solve.

20. The school cafeteria sold 583 tuna sandwiches, 108 chicken sandwiches, and 223 peanut butter sandwiches in one week. How many sandwiches did it sell in all?

____

21. One month, the school cafeteria sold 1,118 cartons of milk and 2,884 bottles of juice. How many drinks did it sell in all?

____

Name ____________________

Relate Addition and Subtraction • Algebra

Write a fact family for each group of numbers.

1. 8, 9, 17

____

____

____

____

2. 5, 0, 5

____

____

____

____

3. 8, 8, 16

____

____

Find each sum or difference. Write a related addition or subtraction sentence from the same fact family.

4. $13 - 5 =$ ______

____

5. $7 + 4 =$ ______

____

6. $14 - 8 =$ ______

____

7. $12 - 6 =$ ______

____

8. $7 + 8 =$ ______

____

9. $8 - 0 =$ ______

____

Find each missing addend.

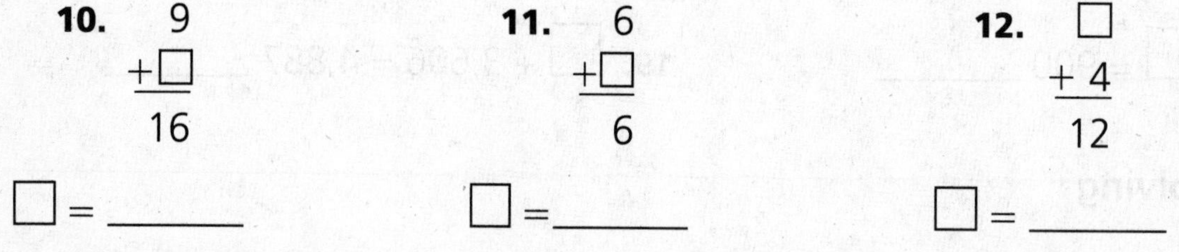

10. $9 + \square = 16$

$\square =$ ______

11. $6 + \square = 6$

$\square =$ ______

12. $\square + 4 = 12$

$\square =$ ______

Problem Solving

Solve.

13. Joe plants 6 tomato plants and 15 pepper plants. How many more pepper plants does he plant?

____

14. Joe picks 4 beefsteak tomatoes and 9 cherry tomatoes. How many tomatoes does he pick in all?

____

Name ____

Problem Solving: Skill
Identify Extra Information

Solve. Cross out the extra information in each problem.

1. Tracy sells $86 worth of flowers. Nick sells $65 worth of flowers. They work at the flower stand from 3:00 P.M. to 7:00 P.M. How much money did Tracy and Nick make in all?

____

2. Elena brought 20 packs of seeds. She bought 14 packs of flower seeds. The rest were vegetable seeds. Five packs held seeds for wild flowers. How many packs of vegetable seeds did Elena buy?

____

3. There are 24 displays at the flower show. One display has 19 flowers. This display has 8 yellow flowers. The rest of the flowers in this display are pink. How many pink flowers are there?

____

4. Grace has 17 bunches of flowers. She sells 5 of them. The flowers cost $10 per bunch. How many bunches of flowers does Grace have left?

____

5. Ron wins 15 gold medals and 2 silver medals. Sue wins 7 gold medals. How many more gold than silver medals does Ron win?

____

6. Lilies cost $1.00 per pair. Theresa buys 6 lilies, 12 roses, and 3 tulips. How many flowers does Theresa buy in all?

____

7. Kip sells 46 tickets to the flower show. Ellen sells 7 tickets to the flower show. The show runs for 2 days and is open for 10 hours each day. How many more tickets does Kip sell than Ellen?

____

8. There are 37 people signed up to work at the flower show. 4 of those people do not come to the show. The show lasts 3 days. How many people work at the flower show?

____

Name ____________________

Subtraction Patterns • Algebra

Write the number that makes each sentence true.

1. 7 − 5 = _____	**2.** 9 − 4 = _____	**3.** 13 − 6 = _____
70 − 50 = _____	90 − 40 = _____	130 − 60 = _____
700 − 500 = _____	900 − 400 = _____	1,300 − 600 = _____
4. 14 − 9 = _____	**5.** 11 − _____ = 9	**6.** 14 − _____ = 8
140 − _____ = 50	_____ − 20 = 90	140 − 60 = _____
1,400 − 900 = _____	1,100 − _____ = 900	_____ − 600 = 800

Subtract mentally.

7. 130 − 70

8. 1,500 − 700

9. 1,000 − 300

10. 1,800 − 900

11. 500 − 500

12. 90 − 60 = _____	**13.** 120 − 50 = _____	**14.** 1,100 − 400 = _____
15. 500 − 100 = _____	**16.** 170 − 90 = _____	**17.** 1,500 − 600 = _____
18. 800 − 500 = _____	**19.** 130 − 70 = _____	**20.** 90 − 90 = _____

Find the missing digit.

21. 400 − ___00 = 200 **22.** 989 − ___00 = 489 **23.** 2,___50 − 600 = 2,350

Problem Solving

Solve.

24. There are 1,000 tickets to the Tulip Show. 400 tickets have been sold. How many tickets are still left?

____

25. Last year, the Tulip Show had 200 kinds of tulips for sale. This year, it has 400 kinds for sale. How many more kinds of tulips are for sale this year?

____

Name ____________________

Explore Regrouping in Subtraction

Use models to subtract.

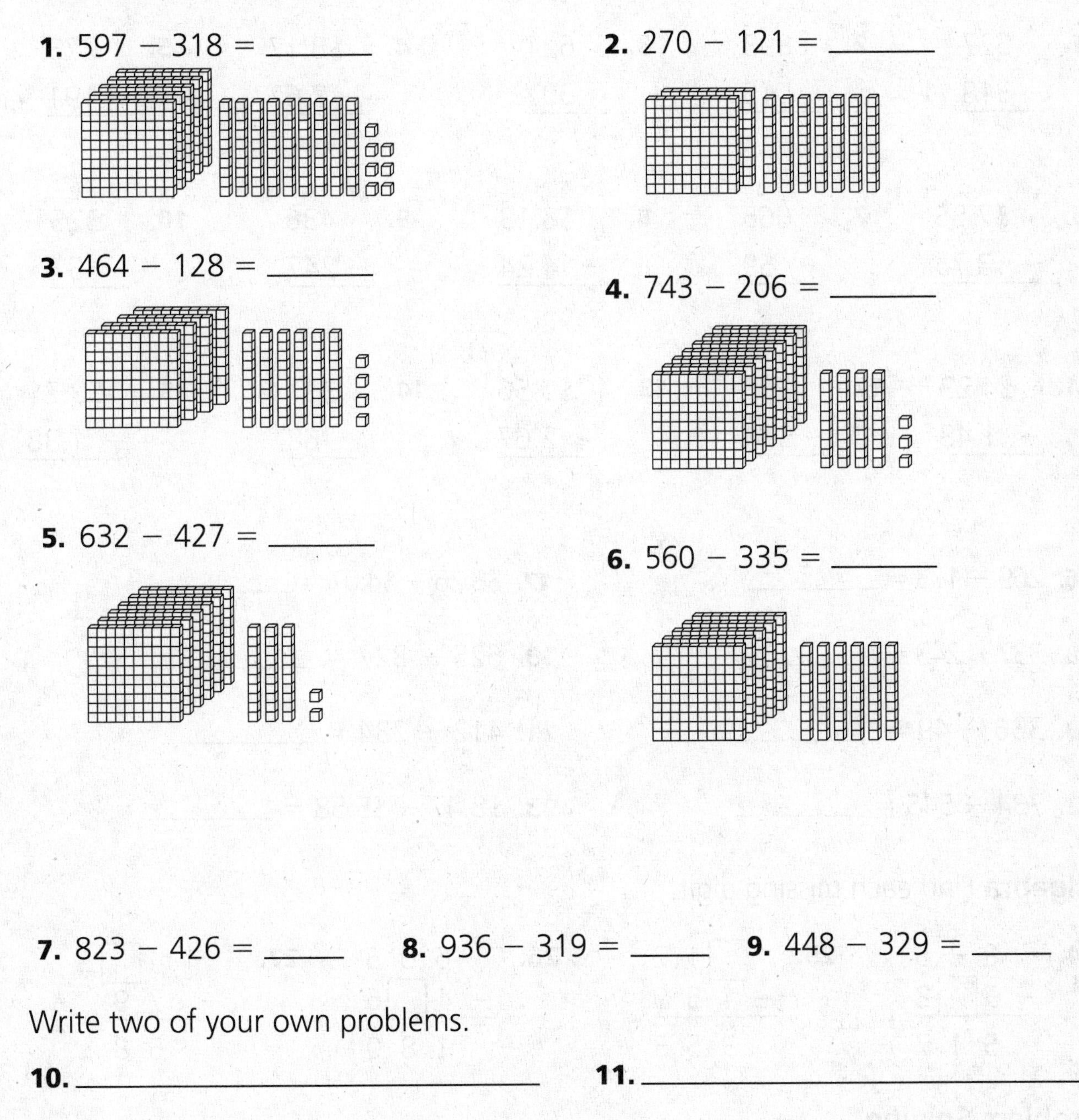

7. 823 − 426 = _____ **8.** 936 − 319 = _____ **9.** 448 − 329 = _____

Write two of your own problems.

10. ______________________________ **11.** ______________________________

Name ______________________________

Subtract Whole Numbers

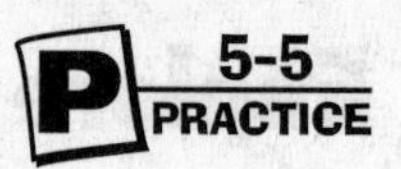

Subtract. Check your answer.

1. $\begin{array}{r} 527 \\ -\ 318 \\ \hline \end{array}$
2. $\begin{array}{r} 893 \\ -\ 515 \\ \hline \end{array}$
3. $\begin{array}{r} 621 \\ -\ 307 \\ \hline \end{array}$
4. $\begin{array}{r} \$8.17 \\ -\ 3.62 \\ \hline \end{array}$
5. $\begin{array}{r} 273 \\ -\ 191 \\ \hline \end{array}$
6. $\begin{array}{r} \$7.55 \\ -\ 3.75 \\ \hline \end{array}$
7. $\begin{array}{r} 665 \\ -\ 59 \\ \hline \end{array}$
8. $\begin{array}{r} \$8.13 \\ -\ 4.24 \\ \hline \end{array}$
9. $\begin{array}{r} 436 \\ -\ 247 \\ \hline \end{array}$
10. $\begin{array}{r} 325 \\ -\ 57 \\ \hline \end{array}$
11. $\begin{array}{r} \$4.23 \\ -\ 1.48 \\ \hline \end{array}$
12. $\begin{array}{r} 763 \\ -\ 574 \\ \hline \end{array}$
13. $\begin{array}{r} \$5.56 \\ -\ 2.67 \\ \hline \end{array}$
14. $\begin{array}{r} 337 \\ -\ 68 \\ \hline \end{array}$
15. $\begin{array}{r} \$3.21 \\ -\ 1.38 \\ \hline \end{array}$

16. 269 − 173 = ________

17. $8.26 − $4.07 = ________

18. 337 − 243 = ________

19. 625 − 327 = ________

20. 336 − 49 = ________

21. 412 − 234 = ________

22. 734 − 545 = ________

23. $8.47 − $5.58 = ________

Algebra Find each missing digit.

24. $\begin{array}{r} 8\ 3\ 5 \\ -\ 3\ \square\ 8 \\ \hline 5\ 1\ 7 \end{array}$
25. $\begin{array}{r} \square\ 1\ 2 \\ -\ 1\ 5\ 9 \\ \hline 3\ 5\ 3 \end{array}$
26. $\begin{array}{r} 6\ 6\ 6 \\ -\ 4\ \square\ 6 \\ \hline 1\ 8\ 0 \end{array}$
27. $\begin{array}{r} 1\ 4\ \square \\ -\ 7\ 9 \\ \hline 6\ 8 \end{array}$

Problem Solving

Solve.

28. Marco needs a total of $7.50 for seeds. He has $2.85. How much more does Marco need?

____

29. Angela has 145 plants in her garden. Joy has 79 plants in her garden. How many more plants does Angela have?

____

Name ______________________________

Regroup Across Zeros

6-1 PRACTICE

Subtract. Add to check.

1. 503 − 82

2. $607 − 238

3. 730 − 467

4. 901 − 719

5. $309 − 223

6. 208 − 75

7. 305 − 161

8. 400 − 286

9. 504 − 316

10. $700 − 199

11. 103 − 45 = ______

12. $901 − $333 = ______

13. 800 − 65 = ______

14. 702 − 618 = ______

15. 205 − 74 = ______

16. 700 − 412 = ______

17. 607 − 31 = ______

18. 800 − 433 = ______

Algebra Use the rule to find the difference.

19. Rule: Subtract 37

Input	Output
200	
150	
600	

20. Rule: Subtract 239

Input	Output
500	
900	
800	

21. Rule: Subtract 89

Input	Output
202	
707	
808	

Problem Solving

Solve.

22. A bag holds 300 seeds. Brandon plants 79 of the seeds. How many seeds are left?

____

23. A book about gardening has 504 pages. Amy has read 245 pages so far. How many more pages does she have left to read?

____

Name ______________________________

Estimate Differences

Estimate each difference.

1. 73 − 27 ______
2. 91 − 65 ______
3. 685 − 193 ______
4. 947 − 831 ______
5. 45 − 19 ______
6. 54 − 38 ______
7. 615 − 315 ______
8. 725 − 199 ______
9. 981 − 350 ______
10. 862 − 98 ______
11. 219 − 68 ______
12. 550 − 75 ______
13. 703 − 376 ______
14. 902 − 829 ______
15. 1,709 − 888 ______
16. 7,233 − 999 ______
17. 1,590 − 690 ______
18. 9,031 − 1,786 ______

Algebra Write > or < to make a true sentence.

19. 58 − 27 ◯ 50
20. 92 − 18 ◯ 80
21. 168 − 79 ◯ 100
22. 705 − 280 ◯ 400
23. 932 − 239 ◯ 800
24. 850 − 176 ◯ 800
25. 48 − 27 ◯ 800
26. 650 − 403 ◯ 200

Problem Solving

Solve.

27. A sugar maple tree is 72 feet tall. A pecan tree is 111 feet tall. About how much taller is the pecan tree?

____

28. A ponderosa pine tree is 143 feet tall. It is 89 feet taller than a red juniper tree. About how tall is the red juniper tree?

____

Name ______________________________

Problem Solving: Strategy
Write an Equation

Write a number sentence to solve.

1. A black spruce tree is 32 feet tall. An Engelmann pine tree is 110 feet tall. How much taller is the Engelmann pine than the black spruce?

____

2. A live oak tree is 48 feet tall. A California white oak tree is 42 feet taller. How tall is the California white oak?

____

3. The garden club raises $123 for a community garden. The club spends $78 on supplies. How much money does the garden club have left?

____

4. Maria spends $246 on plants for her garden. She spends $84 on tools. How much money does Maria spend in all?

____

Mixed Strategy Review

Solve. Use any strategy.

5. Nadia's garden has a length of 45 feet and a width of 32 feet. How much longer is the length than the width?

____

Strategy: ______________________________

6. **Science** Fires can burn forests at a rate of up to 10 miles per hour. How many miles can a forest fire travel in 3 hours?

____

Strategy: ______________________________

7. Jason surveys trees in the park. His results are maple, maple, oak, maple, oak, pine, pine, oak, oak, ash, maple, ash, maple, pine. Which kind of tree is the most common in the park?

____

Strategy: ______________________________

8. **Create a problem** that you could write a number sentence to solve. Share it with others.

____

____

____

____

Name ___

Subtract Greater Numbers

Subtract. Check each answer.

1. 6,387 − 192

2. $62.17 − 38.60

3. 9,817 − 2,087

4. 1,754 − 382

5. $34.98 − 25.67

6. 4,891 − 975

7. $31.65 − 16.20

8. 9,315 − 4,928

9. $4,046 − 2,995

10. 6,635 − 669

11. 5,602 − 325

12. $82.50 − 7.66

13. 3,426 − 2,839

14. $51.63 − 38.86

15. 7,546 − 787

16. 4,008 − 3,912 = ___

17. $82.70 − $50.92 = ___

18. 5,123 − 987 = ___

19. 7,654 − 6,666 = ___

20. $4,325 − $998 = ___

21. 6,000 − 85 = ___

22. 6,200 − 5,375 = ___

23. $50.52 − $24.48 = ___

Algebra Write + or − to make a true number sentence.

24. 8,734 ◯ 4,292 = 4,442

25. 687 ◯ 474 = 1,161

26. $81.32 ◯ $9.83 = $91.15

27. 8,225 ◯ 6,334 = 1,891

Problem Solving

Solve.

28. On a parade float, there are 3,000 red roses and 1,850 white roses. How many more red roses are there?

___

29. Of the 4,208 roses on another float, 680 were wilted. How many were not wilted?

___

Name ______________________________

Choose a Computation Method

Subtract. Tell which method you use.

1. 851 − 242 = ______

2. 630 − 472 = ______

3. 900 − 725 = ______

4. 2,351 − 1,128 = ______

5. 7,633 − 3,981 = ______

6. 8,124 − 4,316 = ______

7. 7,295 − 556 = ______

8. 4,010 − 2,397 = ______

9. 8,734 − 444 = ______

Algebra Find the missing number.

10. 630 − ☐ = 210 ______

11. ☐ − 374 = 105 ______

12. 800 − ☐ = 628 ______

13. ☐ − 3,550 = 1,617 ______

14. 2,503 − ☐ = 1,003 ______

15. ☐ − 1,221 = 4,822 ______

Problem Solving

Solve. Use data from the table.

Bus Riders	
Bus Line	**Number of Riders**
Red Line	1,216
Blue Line	1,330
Green Line	1,592

16. How many more people rode the Blue Line bus than the Red Line bus?

____

17. About how many people rode the 3 bus lines all together?

____

18. Which bus line had the fewest number of riders?

____

Name ____

Tell Time

Write each time using A.M. or P.M. Then write one way to read each time.

1. Bed time

Write: ____________

Read: ____________________________

____

2. Off to school

Write: ____________

Read: ____________________________

____

Write the time in two different ways. Include A.M. and P.M.

3. Lunch time

Write: ____________

Write: ____________________________

____

____

4. School is out

Write: ____________

Write: ____________________________

____

____

Algebra Fill in the missing digits on the digital clocks.

5. **6.** **7.** **8.**

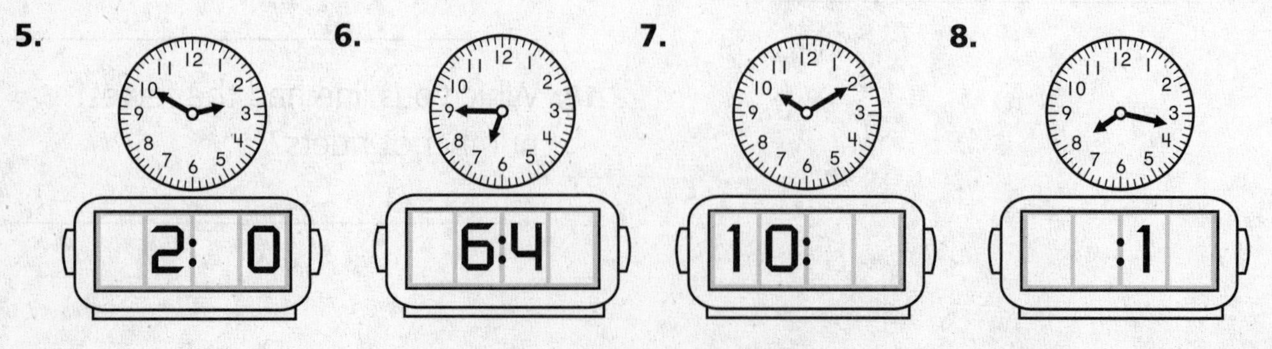

Name ____________________

Convert Time

Tell how much time.

1. 60 minutes = _____ hour

2. $\frac{1}{2}$ hour = _____ minutes

3. 3 hours = _____ minutes

4. 120 minutes = _____ hours

5. $\frac{1}{4}$ hour = _____ minutes

6. 2 half hours = _____ minutes

7. 30 minutes = _____ hour

8. 15 minutes = _____ hour

9. 3 half hours = _____ minutes

10. 3 quarter hours = _____ minutes

11. $1\frac{1}{2}$ hours = _____ minutes

12. 1 hour = _____ quarter hours

13. 2 hours = _____ minutes

14. 4 half hours = _____ hours

15. $\frac{3}{4}$ hour = _____ minutes

16. 30 minutes = _____ quarter hours

17. 2 hours = _____ minutes

18. 150 minutes = _____ half hours

Problem Solving

Solve.

19. Jan exercises for $\frac{1}{2}$ hour. How many minutes is that?

_____ minutes

20. The bus ride is 15 minutes long. How many hours is that?

_____ hour

21. The movie is 2 hours long. How many minutes is that?

_____ minutes

22. Eric trains for 60 minutes. How many quarter hours is that?

_____ quarter hours

Name ___

Elapsed Time

7-3 PRACTICE

How much time has passed?

1. Begin: 4:15 P.M.
End: 4:30 P.M.

___

2. Begin: 7:20 A.M.
End: 8:00 A.M.

___

3. Begin: 11:30 A.M.
End: 1:30 P.M.

___

4. Begin: 9:05 A.M.
End: 5:05 P.M.

___

5. Begin: 10:15 A.M.
End: 11:50 A.M.

___

6. Begin: 9:30 P.M.
End: 6:30 A.M.

___

What time will it be in 3 hours?

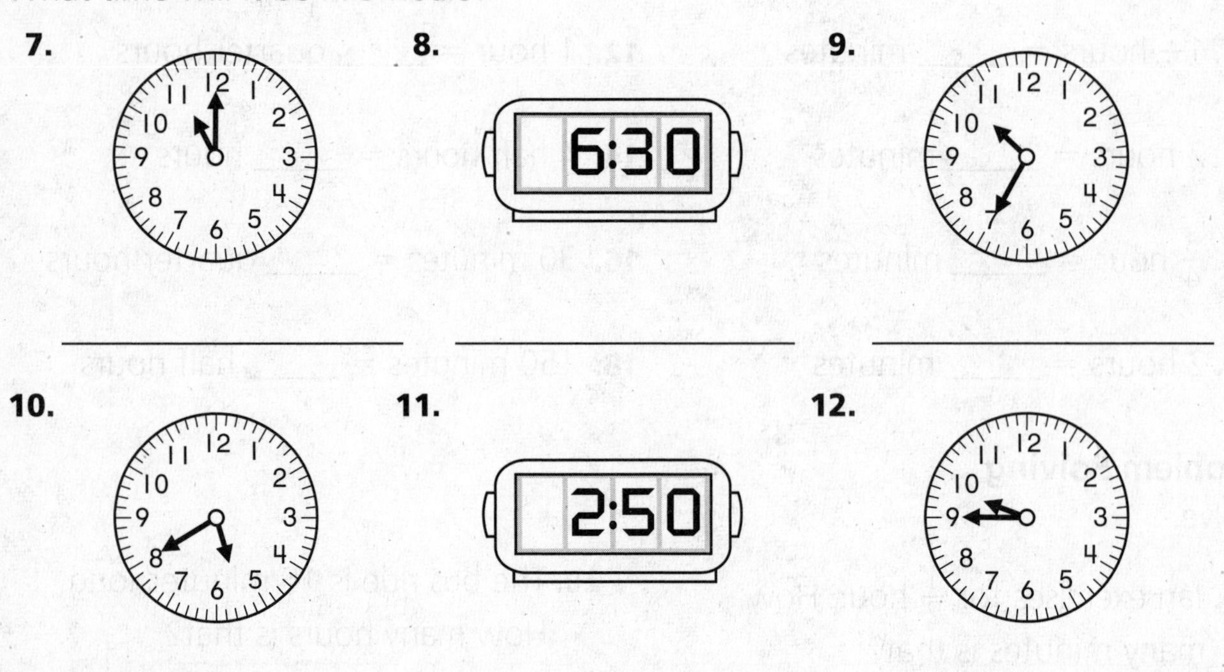

Problem Solving

Solve.

13. Science class begins at 1:10 P.M. and ends at 2:05 P.M. How many minutes long is science class?

___

14. The school day begins at 8:45 A.M. and ends 7 hours later. At what time does the school day end?

___

Name ______________________________

Calendar

May						
S	M	T	W	T	F	S
		1	2	3	4	5
6	7	8	9	10	11	12
13 **Mother's Day**	14	15	16	17	18	19
20	21	22	23 **José's Birthday**	24	25	26
27	28	29	30	31		

Use the calendar to solve each problem.

1. What is the date for the first Thursday in May?

____

2. On which day of the week is Mother's Day?

____

3. On which day of the week is the twenty-eighth?

____

4. How many weeks are there from May 3 through May 24?

____

5. Lana's birthday is on the fourth Saturday in May. What is the date of her birthday?

____

6. How many days are there from José's birthday to May 30th?

____

7. Today is May 4th. José has a soccer game in 2 weeks. What is the date of the soccer game?

____

8. José has a scout meeting on the third Tuesday of every month. What is the date of the scout meeting in May?

____

Name ______________________________

Time Lines

7-5 PRACTICE

Lin went to summer camp. Use Lin's time line to solve problems 1-6.

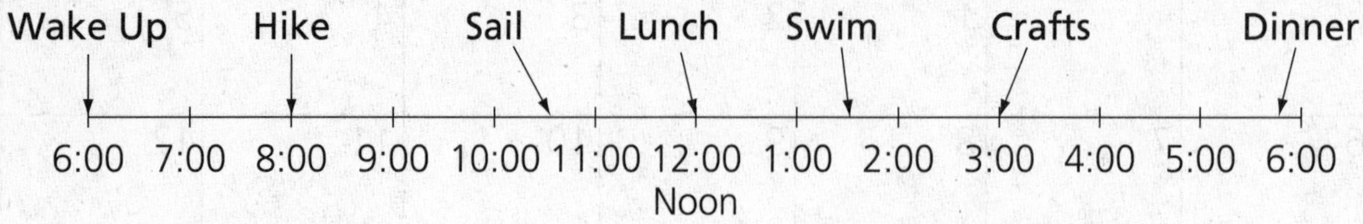

Wake Up (6:00), Hike (8:00), Sail, Lunch (12:00), Swim, Crafts (3:00), Dinner

6:00 7:00 8:00 9:00 10:00 11:00 12:00 Noon 1:00 2:00 3:00 4:00 5:00 6:00

1. What happens at 8:00 A.M.?

____

2. What activity happens between hiking and lunch?

____

3. What time does Lin eat lunch?

____

4. How long is it between sailing and lunch?

____

5. What happens after swimming?

____

6. Breakfast is at 6:30 A.M. Where does it belong on the time line?

____

Problem Solving

Solve.

7. On a time line showing the order that the children in a family were born, Amy is to the right of Jeremy. Randy is to the left of Jeremy. Draw the timeline.

8. Write the order of the children's births from oldest to youngest.

____

Use with Grade 3, Chapter 7, Lesson 5, pages 152–153.

Name ______________________________

Problem Solving: Skill

Identify Missing Information

Solve. Identify the missing information.

1. Annie is at basketball practice from 2:45 P.M. to 4:15 P.M. Fifteen minutes after basketball practice ends, Annie gets home and starts her homework. She does homework for 45 minutes. At what time does Annie finish her homework?

____

____

2. Allie goes to visit a friend at 3:00 P.M. She stays for 1 hour and then walks to the library. She gets to the library at 4:20. How long is the walk from her friend's house to the library?

____

____

3. Brandy ate dinner from 7:00 P.M. to 7:45 P.M. Then she read until 8:30 P.M. After that, she watched television until 9:00 P.M. How long did Brandy read?

____

____

4. Miguel goes to a 2 hour-long movie that starts at 3:30 P.M. Right after the movie ends, Miguel walks home. Miguel gets home at 6:25 P.M. How long does the walk take?

____

____

Mixed Strategy Review

5. Sam leaves school at 3:15. He gets to the mall at 3:45 P.M. and spends 45 minutes there. Then he walks home and arrives there at 5:30 P.M. How many hours after he left school did Sam arrive home?

____

6. Tony practices the drums from 12:00 P.M. to 12:30 P.M. Then he reads for 45 minutes. Immediately after that, Lou practices the drums for 40 minutes. At what time did he finish practicing the drums?

____

Name ______________________________

Collect and Organize Data

Stan surveyed third-grade students to find the number of different states they have visited. First he recorded the responses in a tally chart. Then he began to make a line plot.

1. Complete Stan's tally chart at right. Write the number of responses in the last column.

How Many States Have You Visited?		
Number of States Visited	Tally	Number of Responses
1	////	4
2	~~////~~ ~~////~~	
3	~~////~~ ///	
4	///	
5 or more	//	

2. How many third-grade students did Stan survey?

____

3. How many third-grade students have visited 5 or more different states?

____

4. How many third-grade students have visited fewer than 3 different states?

____

5. Finish Stan's line plot.

6. How many different states have the most number of responses?

____

7. Stan surveyed 3 more third-grade students. Each of them has visited 3 different states. How does this change the tally chart and the line plot?

____

____

____

How Many States Have You Visited?

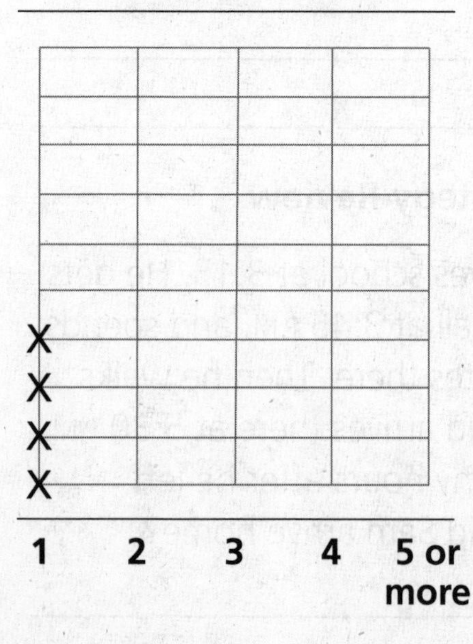

Name ______________________________

Find Median, Mode, and Range

Find the range, median, and mode in exercises 1–4.

1. 4, 6, 5, 4, 3, 3, 4, 3, 5

____

2. 10, 12, 14, 10, 13, 10, 13, 11, 13, 14, 11

____

3. 25, 20, 20, 40, 30, 20, 35, 25, 20, 40, 30

____

4. 9, 8, 10, 7, 11, 11, 10, 7, 9, 11, 7

____

Use the data from the tally chart.

Number of Loons Seen	
Week	**Tally**
1	/
2	////
3	///
4	///

5. What is the range of the data?

____

6. What is the median of the data?

____

7. What is the mode of the data?

____

Problem Solving

Solve.

8. Mandy found the range, median, and mode of 5 numbers. The range was 2, the median was 2, and the mode of the numbers was 2. What were the 5 numbers?

____

____

Name ____

Explore Pictographs

Mr. Mendoza surveyed some students at Roosevelt Elementary School to find out what types of stories they like to read. First he made a chart of their responses. Then he started to make a pictograph.

Favorite Types of Stories

Type of Story	Number of Students
Adventure	30
Animal	40
Space	15
Fantasy	25

Favorite Types of Stories

Adventure	📖 📖 📖 📖 📖 📖
Animal	📖 📖 📖 📖 📖 📖 📖 📖
Space	📖 📖 📖
Fantasy	📖 📖 📖 📖 📖

Key: Each 📖 stands for 5 responses.

Use the data from the pictograph. Then answer each question.

1. Look at the pictograph. What does each 📖 stand for?

____

____

2. How many 📖 are shown for adventure stories in the pictograph? Tell why.

____

____

3. Which type of story is the students' favorite?

____

4. How many more students like adventure stories than space stories?

____

5. How many students were in the survey? How can you tell?

____

____

6. Suppose 20 students liked to read space stories. How would the pictograph change?

____

____

Name ____

Explore Bar Graphs

Yoshi finds the following data about the life span of some animals. First he records the data in a chart. Then he starts to make a bar graph.

Average Animal Life Span	
Type of Animal	Average Life Span (Years)
Black bear	18
Domestic cat	12
Chipmunk	6
Domestic dog	12
Cow	15

Average Animal Life Span

Number of Years: 0, 2, 4, 6, 8, 10, 12, 14, 16, 18, 20

Animal: Bear, Cat, Chipmunk, Dog, Cow

Use the data in the chart to finish the bar graph. Then answer each problem.

1. How did you know how to show the number of cows even though 15 is not listed on the scale of the graph?

____

____

2. Why does the graph show every second number instead of all the numbers from 0 through 20 in the scale?

____

____

3. Which animal has the longest average life span? ____________

4. Which animals have about the same average life span? ____________

5. How many more years is a cow more likely to live than a cat? ____________

6. Which animal has the shortest life span? ____________

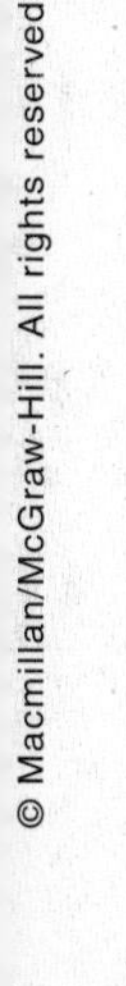

Name ____________________

Problem Solving: Strategy

Work Backward

Work backward to solve.

1. Wendy finishes practice at 4:15 P.M. The practice was 45 minutes. At what time did practice start?

____

2. Paula has $4 left. She spent $5 on a book and $3 on a snack. How much money did Paula start with?

____

3. Ben buys a camera with a fifty-dollar bill. With his change, he buys a photo frame for $12. Now he has $8. How much was the camera?

____

4. Nick leaves his house and rides his bike to the mall. He spends 1 hour at the mall. Nick is back home at 5:00 P.M. The bike ride is 30 minutes each way. At what time did Nick leave his house to go to the mall?

____

Mixed Strategy Review

Solve. Use any strategy.

5. Juan leaves the library. He walks 15 minutes and stops at a store for 15 minutes. Juan leaves the store at 10:15 A.M. At what time did Juan leave the library?

____

Strategy: ____________________

6. On Wednesday, 4,267 people attended the music fair. On Thursday, 3,385 people attended the fair. How many people attended the fair in all?

____

Strategy: ____________________

7. **Health** You can estimate your maximum heart rate by subtracting your age (in years) from 220. Val estimates her maximum heart rate as 208. How old is Val?

____

Strategy: ____________________

8. **Write a problem** which can be solved by working backward. Share it with others.

____

____

____

Use with Grade 3, Chapter 8, Lesson 5, pages 170–171.

Name ____________________

Coordinate Graphs

Use the map to write the location.

1. Fire station ________

2. Jacob's house ________

3. Video store ________

4. Public library ________

5. School ________

6. Hospital ________

Jacob's House
Hospital
Bike Store
Grocery Store
Video Store
Deli
Fire Station
School
Public Library

6
5
4
3
2
1
0
0 1 2 3 4 5 6

Name the place at each location on the map.

7. (5,1) ________

8. (3, 4) ________

9. (1, 3) ________

10. (4, 3) ________

11. (0,0) ________

12. (2, 2) ________

Problem Solving

Solve.

13. Jacob's friend Martin lives nearby. The location of his house is (4, 5). Draw Martin's house on the map above.

14. The post office is located between the school and the public library. Write the ordered pair for the location of the post office.

____

____

Name ____________________

Interpret Line Graphs

The third-grade class collects bottles each week to earn money for a class trip. The students record data for five weeks of collecting.

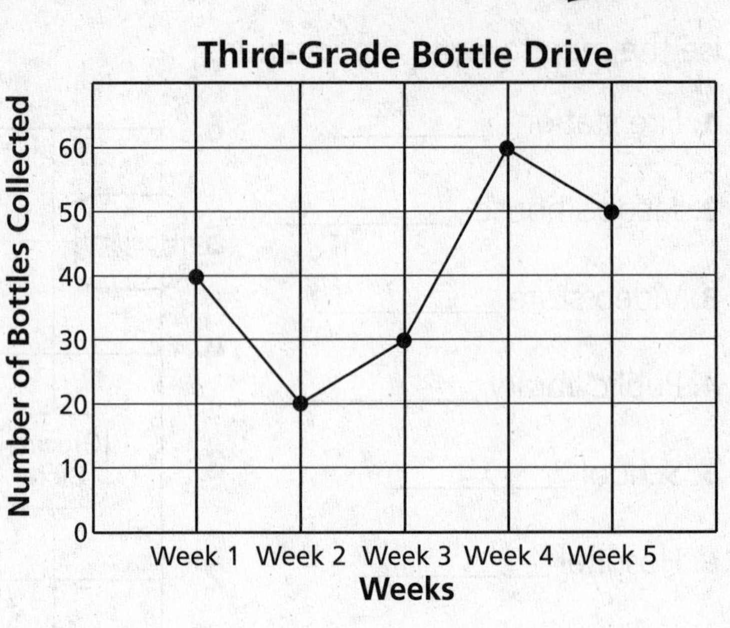

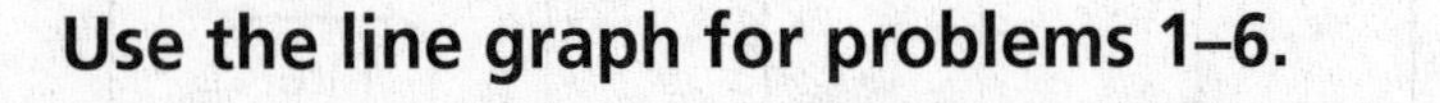

Use the line graph for problems 1–6.

1. How many cans were collected during Week 1?

____

2. During which week were 50 cans collected?

____

3. Between which two weeks did the number of cans collected increase by 10?

____

4. During which week was the greatest number of cans collected?

____

Problem Solving

Solve.

5. How many more cans did the students collect in Week 4 than in the week they collected the least number of cans?

____

6. For which two weeks did they collect fewer cans than they had the week before?

____

Name ____________________

Explore the Meaning of Multiplication

P 9-1 PRACTICE

Use models to find the total number.

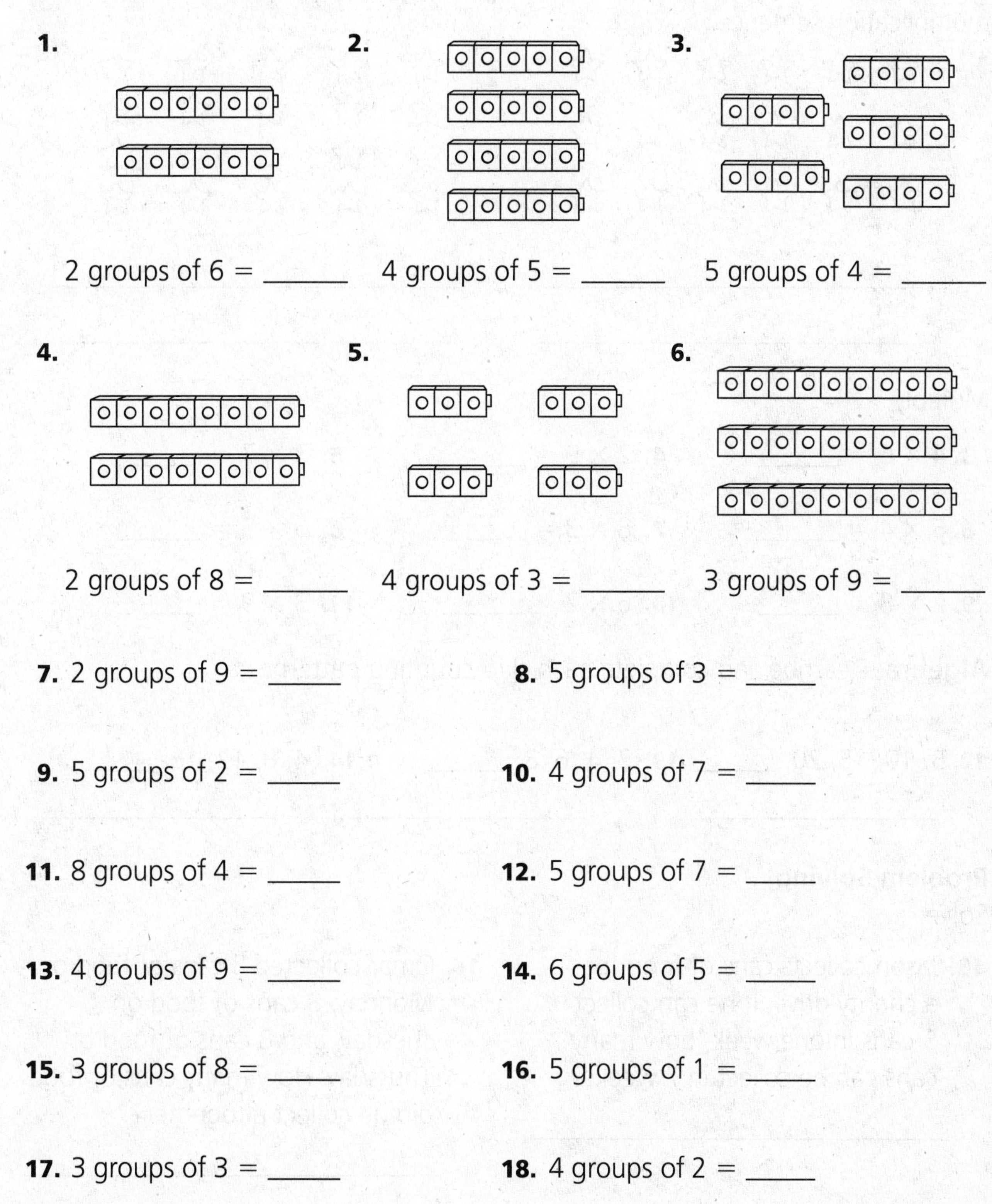

1. 2 groups of 6 = ______

2. 4 groups of 5 = ______

3. 5 groups of 4 = ______

4. 2 groups of 8 = ______

5. 4 groups of 3 = ______

6. 3 groups of 9 = ______

7. 2 groups of 9 = _____

8. 5 groups of 3 = _____

9. 5 groups of 2 = _____

10. 4 groups of 7 = _____

11. 8 groups of 4 = _____

12. 5 groups of 7 = _____

13. 4 groups of 9 = _____

14. 6 groups of 5 = _____

15. 3 groups of 8 = _____

16. 5 groups of 1 = _____

17. 3 groups of 3 = _____

18. 4 groups of 2 = _____

Name ______________________________

Relate Multiplication and Addition • Algebra

Find each total. Write an addition sentence and a multiplication sentence.

1. ☆ ☆ ☆ ☆ ☆ ☆
☆ ☆ ☆ ☆ ☆ ☆
☆ ☆ ☆ ☆ ☆ ☆

____

____

2. ☆ ☆ ☆ ☆ ☆
☆ ☆ ☆ ☆ ☆
☆ ☆ ☆ ☆ ☆
☆ ☆ ☆ ☆ ☆

____

____

Multiply.

3. 4 × 6 = ______ **4.** 2 × 9 = ______ **5.** 3 × 7 = ______

6. 6 × 4 = ______ **7.** 8 × 3 = ______ **8.** 5 × 5 = ______

9. 2 × 8 = ______ **10.** 6 × 2 = ______ **11.** 3 × 9 = ______

Algebra Describe and complete each skip-counting pattern.

12. 5, 10, 15, 20, ______ **13.** 2, 4, 6, 8, ______ **14.** 4, 8, 12, 16, ______

______________ ______________ ______________

Problem Solving

Solve.

15. Jason collects cans of food for a charity drive. If he can collect 5 cans in one week, how many cans can he collect in 7 weeks?

____

16. Omar collected 8 cans of food on Monday, 8 cans of food on Tuesday, and 8 cans of food on Thursday. How many cans of food did he collect altogether?

____

Name ______________________________

Explore Multiplication Using Arrays

Write the multiplication sentence that each array shows.

1.

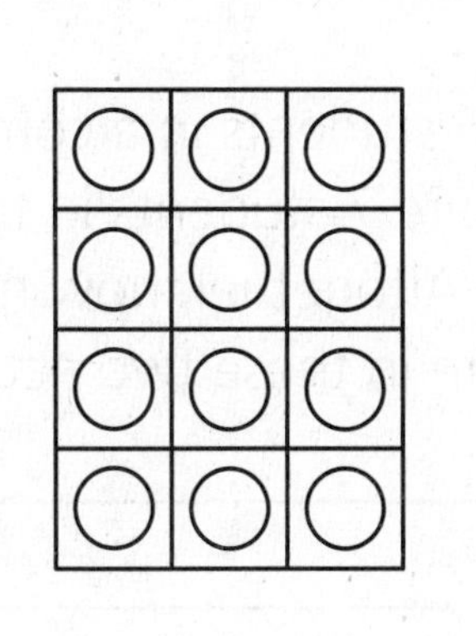

____

2.

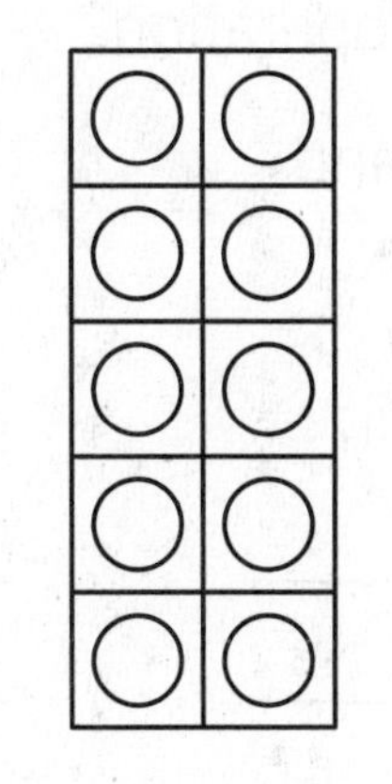

____

3.

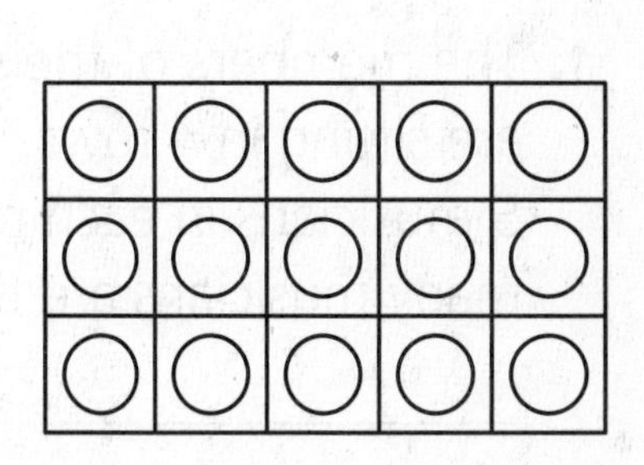

____

Find each product.

4. 2 × 3 = ______ **5.** 3 × 4 = ______ **6.** 5 × 4 = ______

7. 3 × 3 = ______ **8.** 4 × 8 = ______ **9.** 6 × 6 = ______

10. 4 × 9 = ______ **11.** 2 × 7 = ______ **12.** 2 × 5 = ______

13. 3 × 1 = ______ **14.** 8 × 3 = ______ **15.** 5 × 0 = ______

16. 6 × 7 = ______ **17.** 6 × 5 = ______ **18.** 8 × 8 = ______

19. 7 × 4 = ______ **20.** 1 × 9 = ______ **21.** 8 × 6 = ______

22. 7 × 5 = ______ **23.** 9 × 4 = ______ **24.** 5 × 9 = ______

Find each product. Then use the Commutative Property to write a different multiplication sentence.

25. 5 × 3 = ______ **26.** 4 × 6 = ______ **27.** 6 × 1 = ______

______________ ______________ ______________

28. 3 × 8 = ______ **29.** 2 × 9 = ______ **30.** 5 × 6 = ______

______________ ______________ ______________

Name ____________________

Problem Solving: Skill
Choose an Operation

Solve. Explain why you chose the operation.

1. The members of the school band are sitting in 4 rows. There are 5 musicians in each row. How many musicians are there in all?

____
____
____
____

2. There are 6 students in a comedy act. There are 7 students in the dance act. Altogether, how many students are in these two acts?

____
____
____
____

3. A dance act has 3 groups of dancers. There are 6 dancers in each group. How many dancers are in the act?

____
____
____
____

4. There are 3 rows of dancers. Each row has 5 dancers. How many dancers are there in all?

____
____
____
____

Mixed Strategy Review

5. There are 16 children in a comedy act. There are 14 children in a musical act. How many more students are in the comedy act than in the musical act?

____
____
____
____

6. On Friday, 116 people went to the school play. On Saturday, 138 people went to the school play. How many people went to the school play on Friday and Saturday?

____
____
____

Name ______

Multiply by 2 and 5

10-1 PRACTICE

Find each product.

1. $\begin{array}{r} 7 \\ \times 2 \\ \hline \end{array}$
2. $\begin{array}{r} 9 \\ \times 2 \\ \hline \end{array}$
3. $\begin{array}{r} 8 \\ \times 5 \\ \hline \end{array}$
4. $\begin{array}{r} 7 \\ \times 2 \\ \hline \end{array}$
5. $\begin{array}{r} 5 \\ \times 9 \\ \hline \end{array}$
6. $\begin{array}{r} 5 \\ \times 1 \\ \hline \end{array}$

7. $3 \times 2 =$ ______
8. $3 \times 5 =$ ______
9. $5 \times 2 =$ ______
10. $5 \times 8 =$ ______
11. $9 \times 2 =$ ______
12. $9 \times 5 =$ ______
13. $2 \times 8 =$ ______
14. $7 \times 5 =$ ______
15. $2 \times 4 =$ ______
16. $6 \times 2 =$ ______
17. $7 \times 2 =$ ______
18. $8 \times 5 =$ ______
19. $2 \times 2 =$ ______
20. $5 \times 6 =$ ______
21. $4 \times 2 =$ ______
22. $7 \times 5 =$ ______
23. $8 \times 2 =$ ______
24. $5 \times 5 =$ ______

Algebra & Functions Complete the function tables.

25.

Rule: Multiply by 2					
Input	5			8	9
Output	10	12	14		

26.

Rule: Multiply by 5					
Input	4	5		7	
Output	20		30		40

Problem Solving

Solve.

27. The dancers in a ballet class rehearse for 3 hours each day. For how many hours will they rehearse from Tuesday through Saturday?

______

28. The beginner ballet class meets for 6 weeks Tuesday through Saturday. For how many days does the ballet class meet?

______

Name ______________________________

Problem Solving: Strategy

Find a Pattern

Find a pattern to solve.

1. A dancer practices 3 days in a row and then takes one day off to rest. She has a show in two weeks. If she practices on the first 3 days, and takes the 4th day off, how many times will she practice in 14 days?

____

2. The concert hall offers specials on tickets. When you buy 5 tickets, you get 1 other ticket free. When you buy 10 tickets, you get 2 other tickets free. Lyddie got 4 tickets free. How many tickets did she buy?

____

3. The marching band lines up in rows. The first row has 2 people. The second row has 4 people. The third row has 6 people. If this pattern continues, how many people will be in row 5?

____

4. The Portsmouth Players give 2 daytime shows and 3 evening shows per week. Their current play will run for 30 shows. How many of the shows will be daytime shows?

____

Mixed Strategy Review

5. Ken takes piano lessons. The 1st week, he practices 20 minutes each day. The 2nd week, he practices 40 minutes each day. The 3rd week, he practices 1 hour each day. If this pattern continues, how many minutes will he practice each day in the 5th week?

____

6. A theatre seat in the orchestra costs $32. A balcony seat costs $14. How much more does an orchestra seat cost than a balcony seat?

____

Name ______________________________

Multiply by 3 and 4

Multiply.

1. $\begin{array}{r} 1 \\ \times 3 \\ \hline \end{array}$

2. $\begin{array}{r} 8 \\ \times 4 \\ \hline \end{array}$

3. $\begin{array}{r} 4 \\ \times 2 \\ \hline \end{array}$

4. $\begin{array}{r} 3 \\ \times 2 \\ \hline \end{array}$

5. $\begin{array}{r} 4 \\ \times 9 \\ \hline \end{array}$

6. $\begin{array}{r} 3 \\ \times 6 \\ \hline \end{array}$

7. $3 \times 2 =$ ______
8. $4 \times 6 =$ ______
9. $4 \times 4 =$ ______
10. $3 \times 8 =$ ______
11. $3 \times 5 =$ ______
12. $3 \times 7 =$ ______
13. $2 \times 4 =$ ______
14. $3 \times 6 =$ ______
15. $4 \times 5 =$ ______
16. $4 \times 7 =$ ______
17. $3 \times 9 =$ ______
18. $4 \times 8 =$ ______
19. $3 \times 3 =$ ______
20. $4 \times 3 =$ ______
21. $8 \times 4 =$ ______
22. $4 \times 9 =$ ______
23. $3 \times 4 =$ ______
24. $7 \times 3 =$ ______

Algebra Complete.

25.

Number of Boxes	Number of Snacks
3	9
4	12
5	
	18
7	

26.

Number of Packs	Number of Juice Boxes
3	12
4	16
	20
6	
7	

Problem Solving

Solve.

27. There are 4 rows of 9 chairs in the room. How many chairs are in the room?

____

28. There are 3 rows of 7 students in a class photograph. How many students are in the photograph?

____

Name ____________________

Multiply by 0 and 1 • Algebra

Find each product.

1. $\begin{array}{r} 5 \\ \times 1 \\ \hline \end{array}$ 2. $\begin{array}{r} 3 \\ \times 0 \\ \hline \end{array}$ 3. $\begin{array}{r} 8 \\ \times 1 \\ \hline \end{array}$ 4. $\begin{array}{r} 1 \\ \times 7 \\ \hline \end{array}$ 5. $\begin{array}{r} 0 \\ \times 1 \\ \hline \end{array}$ 6. $\begin{array}{r} 1 \\ \times 8 \\ \hline \end{array}$

7. $0 \times 5 =$ ______ 8. $9 \times 0 =$ ______ 9. $0 \times 4 =$ ______

10. $1 \times 4 =$ ______ 11. $1 \times 2 =$ ______ 12. $9 \times 1 =$ ______

13. $1 \times 6 =$ ______ 14. $7 \times 1 =$ ______ 15. $1 \times 3 =$ ______

16. $6 \times 0 =$ ______ 17. $0 \times 2 =$ ______ 18. $5 \times 1 =$ ______

Find each missing number.

19. $6 \times$ ______ $= 6$ 20. ______ $\times 9 = 0$ 21. $1 \times$ ______ $= 1$

22. ______ $\times 7 = 0$ 23. $5 \times$ ______ $= 5$ 24. ______ $\times 4 = 0$

25. $8 \times$ ______ $= 8$ 26. ______ $\times 3 = 0$ 27. $2 \times$ ______ $= 0$

Problem Solving

Solve.

28. There is 1 row of 7 chairs in the back of the classroom. How many chairs are there?

____

29. There are 6 chairs around the table but no one is sitting on them. How many people are sitting on the chairs?

____

Name ________________________________

Explore Square Numbers

Write a multiplication sentence.

1. 2. 3.

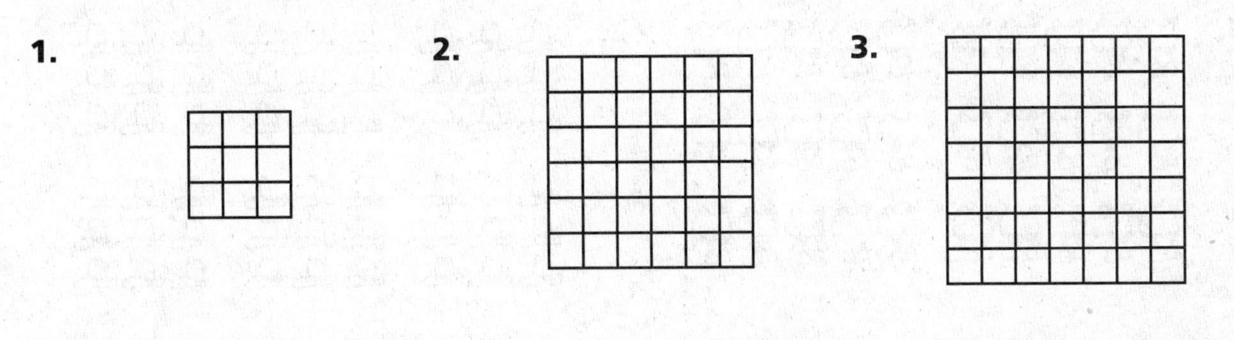

______ ______ ______

Draw a square model and find the product.

4. $4 \times 4 =$ ______ **5.** $8 \times 8 =$ ______ **6.** $9 \times 9 =$ ______

Multiply. You may use grids.

7. $2 \times 2 =$ ______ **8.** $5 \times 5 =$ ______ **9.** $1 \times 1 =$ ______

10. $7 \times 7 =$ ______ **11.** $10 \times 10 =$ ______ **12.** $8 \times 8 =$ ______

Problem Solving

Solve.

13. Myra drew a square on a grid. It was 7 squares long and 7 squares wide. How many small squares were inside the large square?

____

14. Kyle laid some square clay tiles in the garden for a sitting area. How many tiles were laid in an area that is 9 tiles long and 9 tiles wide?

____

Name ______________________________

Multiply by 6 and 8

11-2 PRACTICE

Write a multiplication sentence for the picture.

1. ______________________

2. ______________________

3. ______________________

4. ______________________

Find each product.

5. 6×4

6. 8×8

7. 8×4

8. 6×8

9. 8×9

10. 2×8

11. $5 \times 8 =$ ______

12. $6 \times 3 =$ ______

13. $6 \times 6 =$ ______

14. $6 \times 7 =$ ______

15. $8 \times 4 =$ ______

16. $8 \times 1 =$ ______

17. $8 \times 8 =$ ______

18. $6 \times 9 =$ ______

19. $8 \times 7 =$ ______

Problem Solving

Solve.

20. There are 6 groups of horseback riders on the trail. Each group has 7 riders. How many riders are there in all?

____

21. Each of 8 tourists bought a woven Native American basket that costs $9. How much money did the 8 baskets cost in all?

____

Name ____

Multiply by 7

Write multiplication sentences.

1. How many signs?

____

2. How many fingers?

____

Find each product.

3. 7×3

4. 7×6

5. 7×9

6. 7×7

7. 5×7

8. 2×7

9. $7 \times 4 =$ ______

10. $7 \times 5 =$ ______

11. $7 \times 1 =$ ______

12. $7 \times 8 =$ ______

13. $7 \times 6 =$ ______

14. $7 \times 7 =$ ______

Algebra Complete the table.

15.

Rule: ______________________

Input	Output
5	35
6	42
7	49
8	56

16.

Rule: Multiply by 7

Input	Output
3	
	35
7	
	63

Problem Solving

Solve.

17. Jason read about trains for 2 hours each day for 1 week. How many hours did Jason read?

____

18. Seven people take the train to the next town. The price of a ticket is $9. How much money does the conductor collect?

____

Name ____________________

Find Missing Factors • Algebra

Find the missing factor.

1. ☐ groups of 2 is 18.
2. 3 groups of ☐ is 9.
3. 4 groups of ☐ is 12.
4. ☐ groups of 5 is 10.
5. ☐ groups of 6 is 18.
6. 7 groups of ☐ is 14.
7. $b \times 2 = 12$ ______
8. $4 \times c = 20$ ______
9. $g \times 9 = 9$ ______
10. $6 \times k = 24$ ______
11. $8 \times s = 16$ ______
12. $h \times 7 = 35$ ______

Algebra Complete the table.

13.

Number of Boxes	Number of Books in Each Box	Total Number of Books
2	☐	16
☐	4	20
4	☐	24
7	☐	28
☐	8	32

Problem Solving

Solve.

14. A van can hold 6 passengers. How many trips will the van need to make to give 18 passengers a ride?

______

15. A van made 4 trips. It carried the same number of passengers each trip. It carried 20 passengers in all. How many passengers did it carry each trip?

______

16. The product of a factor and 7 is 21. What is the missing factor?

______

17. The product of 9 and another factor is 63. What is the missing factor?

______

Name ____________________

Problem Solving: Skill
Solve Multistep Problems

Solve.

1. There are 4 rows of seats in the first-class part of a plane. There are 6 rows of seats in the business-class part of the same plane. There are 4 seats in each row. How many seats are there altogether in these sections?

   ____

2. Mr. and Mrs. Lopez take 4 members of the school's theater club to a show. Adult tickets cost \$9 and student tickets cost \$5. How much does the group spend on tickets?

   ____

3. Mr. and Mrs. Jefferson take their 3 children to a Revolutionary War fort. Tickets cost \$7 for adults and \$5 for children. How much do the Jeffersons spend?

   ____

4. Nick buys 6 packs of postcards and 2 packs of souvenir photos. A pack of postcards costs \$4 and a pack of photos costs \$5. How much does Nick spend?

   ____

Mixed Strategy Review

5. Mr. Hong takes a bus to the city. He leaves his house at 8:15 A.M. The bus ride takes 30 minutes. Then he takes a subway to get to a meeting. The subway ride takes 15 minutes. What time does he get to the meeting?

   ____

6. Eight passengers each took 2 suitcases on a plane. 32 passengers each took one suitcase. How many suitcases did the passengers take on the plane in all?

   ____

Name ____________________

Multiply by 10

P 12-1 PRACTICE

Find the product.

1. 10×3 = ____

2. 10×6 = ____

3. 0×10 = ____

4. 10×8 = ____

5. 10×1 = ____

6. 10×4 = ____

7. 2×10 = ____

8. 10×5 = ____

9. 10×7 = ____

10. 10×9 = ____

11. $10 \times 2 =$ ______

12. $10 \times 6 =$ ______

13. $8 \times 10 =$ ______

14. $1 \times 10 =$ ______

15. $4 \times 10 =$ ______

16. $10 \times 7 =$ ______

17. $5 \times 10 =$ ______

18. $3 \times 10 =$ ______

19. $10 \times 0 =$ ______

Problem Solving

Solve.

Use data from the pictograph for problems 20–22.

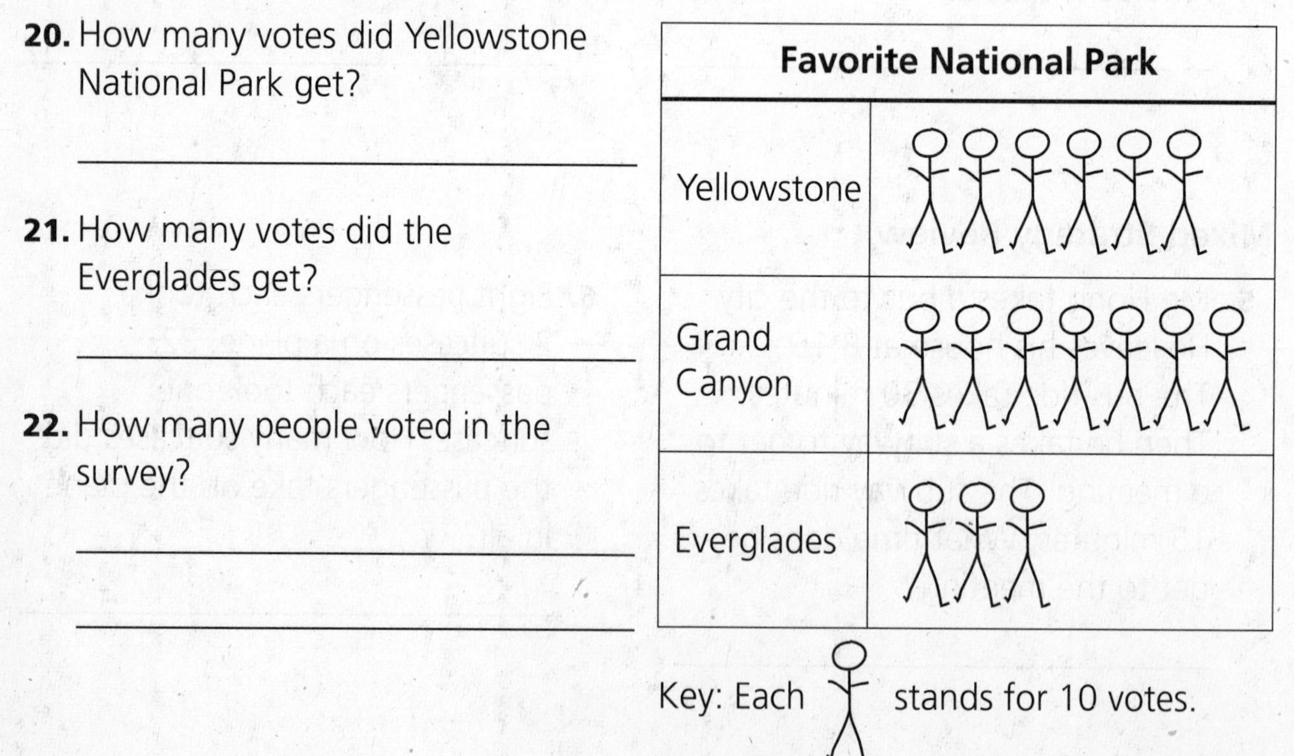

20. How many votes did Yellowstone National Park get?

____

21. How many votes did the Everglades get?

____

22. How many people voted in the survey?

____

____

Name ____________________

Multiply by 9

12-2 PRACTICE

Find each product.

1. $\begin{array}{r} 9 \\ \times 3 \\ \hline \end{array}$ **2.** $\begin{array}{r} 9 \\ \times 8 \\ \hline \end{array}$ **3.** $\begin{array}{r} 4 \\ \times 9 \\ \hline \end{array}$ **4.** $\begin{array}{r} 9 \\ \times 1 \\ \hline \end{array}$ **5.** $\begin{array}{r} 9 \\ \times 7 \\ \hline \end{array}$ **6.** $\begin{array}{r} 9 \\ \times 5 \\ \hline \end{array}$

7. $\begin{array}{r} 9 \\ \times 2 \\ \hline \end{array}$ **8.** $\begin{array}{r} 5 \\ \times 9 \\ \hline \end{array}$ **9.** $\begin{array}{r} 9 \\ \times 0 \\ \hline \end{array}$ **10.** $\begin{array}{r} 9 \\ \times 9 \\ \hline \end{array}$ **11.** $\begin{array}{r} 9 \\ \times 6 \\ \hline \end{array}$ **12.** $\begin{array}{r} 8 \\ \times 9 \\ \hline \end{array}$

13. $2 \times 9 =$ ______ **14.** $4 \times 9 =$ ______ **15.** $9 \times 6 =$ ______

16. $8 \times 9 =$ ______ **17.** $9 \times 1 =$ ______ **18.** $7 \times 9 =$ ______

19. $3 \times 9 =$ ______ **20.** $9 \times 9 =$ ______ **21.** $9 \times 0 =$ ______

22. $3 \times 7 =$ ______ **23.** $8 \times 7 =$ ______ **24.** $6 \times 5 =$ ______

25. $2 \times 8 =$ ______ **26.** $6 \times 7 =$ ______ **27.** $7 \times 4 =$ ______

Problem Solving

Solve.

28. Jordan saw 9 airplanes fly over his house every day last week. How many airplanes did Jordan see last week?

____

29. The Sports Cap Company sent 3 caps to each of the 9 starters on a baseball team. How many caps did the company send?

____

Name ______________________________

Multiplication Table

Complete the multiplication table.

X	0	1	2	3	4	5	6	7	8	9	10	11	12
0	0	0	0	0	0	0	0	0	0				
1	0	1	2		4		6	7				11	
2	0	2	4	6		10		14	16		20		24
3		3		9	12				24			33	
4		4	8				24			36			
5	0				20			35				55	60
6	0				24				48		60		72
7	0			21			42						84
8	0				32					72			96
9	0	9		27						81		99	
10	0	10	20					70				110	
11	0	11	22	33	44							121	
12	0	12	24	36			72					132	144

Use the multiplication table to find each product.

1. $9 \times 11 =$ ______ **2.** $12 \times 6 =$ ______ **3.** $11 \times 10 =$ ______

4. $7 \times 12 =$ ______ **5.** $8 \times 9 =$ ______ **6.** $8 \times 7 =$ ______

7. $9 \times 12 =$ ______ **8.** $11 \times 0 =$ ______ **9.** $12 \times 12 =$ ______

Problem Solving

Solve.

10. There are 12 eggs in one dozen. Eli buys 3 dozen eggs. How many eggs does he buy?

____

11. Julia volunteers for 11 hours a week at the animal shelter. How many hours does she volunteer in 4 weeks?

____

Name ____________________

Multiply 3 Numbers • Algebra

Find each product.

1. $3 \times 3 \times 5 =$ ______
2. $7 \times 1 \times 5 =$ ______
3. $2 \times 3 \times 5 =$ ______
4. $5 \times 0 \times 9 =$ ______
5. $3 \times 5 \times 2 =$ ______
6. $9 \times 1 \times 0 =$ ______
7. $8 \times 2 \times 4 =$ ______
8. $3 \times 2 \times 9 =$ ______
9. $4 \times 4 \times 2 =$ ______
10. $4 \times 5 \times 2 =$ ______
11. $7 \times 2 \times 3 =$ ______
12. $8 \times 5 \times 2 =$ ______
13. $2 \times 2 \times 2 \times 2 =$ ______
14. $2 \times 3 \times 4 \times 5 =$ ______
15. $3 \times 5 \times 0 \times 2 =$ ______
16. $(3 \times 3) \times (2 \times 5) =$ ______
17. $(2 \times 3) \times (2 \times 3) =$ ______
18. $(1 \times 7) \times (7 \times 1) =$ ______

Algebra Find the missing numbers.

19. $(3 \times n) \times 2 = 30$ ______
20. $(4 \times k) \times 5 = 60$ ______
21. $(2 \times p) \times (3 \times 2) = 0$ ______
22. $(8 \times m) \times (2 \times 3) = 48$ ______

Problem Solving

Solve.

23. The students rode in 3 minivans to the museum. Each minivan had 3 rows of seats and 2 passengers in each row. How many passengers were there in all?

____

24. Miss Carter works 2 hours each day for 5 days a week. She is paid $8 per hour. How much money does she earn in one week?

____

Name ____________________

Problem Solving: Strategy

Draw a Picture

Draw a picture to solve.

1. A minibus has 5 rows of seats. Each row has 2 seats. How many people can travel in 5 minibuses?

____

2. Alicia made 3 bead bracelets. She put 4 different types of beads on each bracelet. She used 2 beads of each type. How many beads did she use?

____

3. Eric works on his homework for 2 hours a day. He does his homework 4 days a week. How much time does he spend on his homework in 4 weeks?

____

4. A muffin pan holds 6 muffins. Leilani uses 2 pans for each batch of muffins. How many muffins does she make in 4 batches?

____

Mixed Strategy Review

5. Darin got home at 8:00 P.M. He was traveling for 1 hour and 15 minutes. At what time did he start traveling?

____

6. Karen spent $325 for a plane ticket and $190 on a rental car. How much money did she spend altogether?

____

Name ______________________________

Explore the Meaning of Division

Find how many counters in each group.

1.

____

2.

____

Find the number of equal groups.

3. Put 2 counters in each group

____

4. Put 4 counters in each group

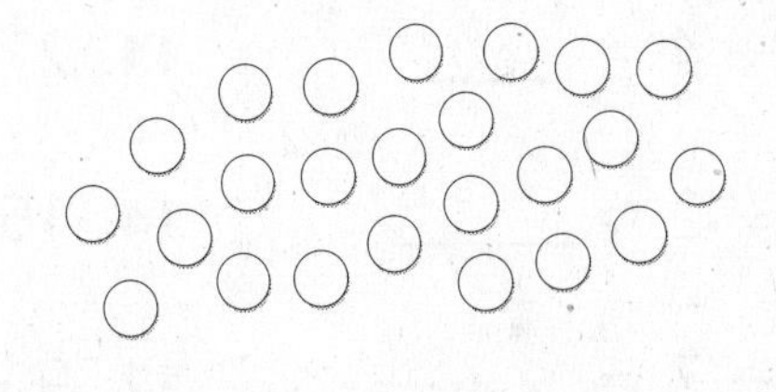

____

Use counters to complete the table.

	Total	Number of Equal Groups	Number in Each Group
5.	18	6	
6.	24		3
7.	32		8
8.	27	3	
9.	36		6

Problem Solving

Solve.

10. Michael has 12 model rocket ships. He puts the same number of models on each of 4 shelves. How many are on each shelf?

____

11. There are 15 astronauts. One shuttle can carry 5 astronauts. How many shuttles are needed?

____

Name ____________________

Relate Division and Subtraction • Algebra

Write how many times you need to subtract.

1. $10 \div 5 =$ ________

2. $21 \div 3 =$ ________

3. $24 \div 4 =$ ________

4. $18 \div 3 =$ ________

5. $30 \div 6 =$ ________

6. $16 \div 2 =$ ________

Divide.

7. $7 \div 7 =$ _____

8. $18 \div 2 =$ _____

9. $28 \div 4 =$ _____

10. $32 \div 4 =$ _____

11. $27 \div 9 =$ _____

12. $36 \div 4 =$ _____

13. $12 \div 2 =$ _____

14. $16 \div 8 =$ _____

15. $25 \div 5 =$ _____

16. $28 \div 7 =$ _____

17. $9 \div 9 =$ _____

18. $20 \div 4 =$ _____

Algebra Complete and describe each pattern.

19. 27, 24, _____, 18, _____

____

20. 48, 42, 36, _____, _____, 18

____

Problem Solving

Solve.

21. In Mr. Ryan's class, 18 students write reports on stars. The students work in groups of 3. Each group hands in 1 report. How many reports on stars did the students hand in?

____

22. Janell pays $20 for 5 astronaut models. Each model costs the same amount. How much does each model cost?

____

Name ______________________________

Relate Multiplication to Division • Algebra

Write related multiplication and division sentences for each picture.

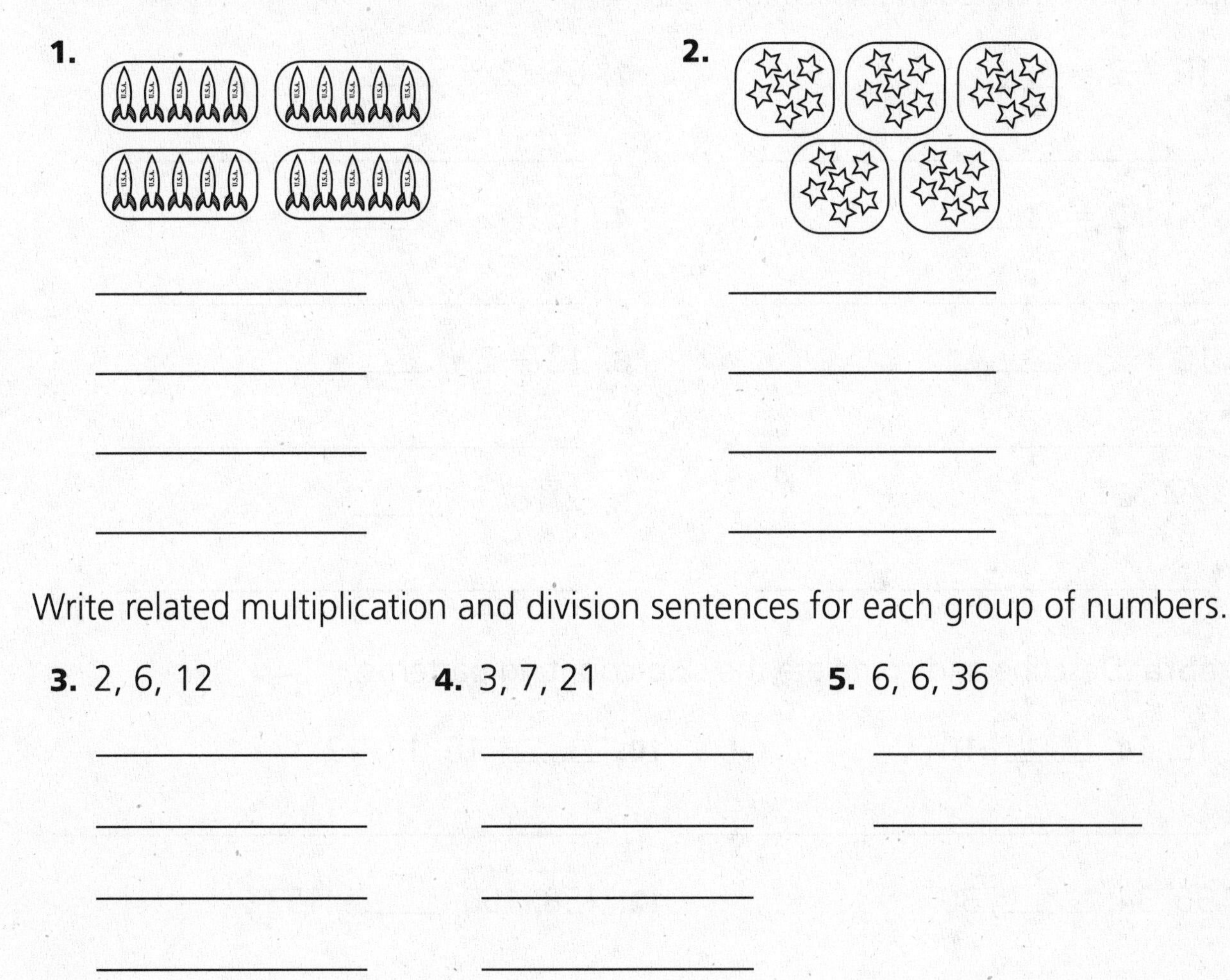

Write related multiplication and division sentences for each group of numbers.

3. 2, 6, 12

4. 3, 7, 21

5. 6, 6, 36

Write × or ÷ to make each sentence true.

6. 54 ◯ 9 = 6

7. 81 ◯ 9 = 9

8. 9 ◯ 5 = 45

9. 8 ◯ 4 = 32

10. 16 ◯ 8 = 2

11. 10 ◯ 1 = 10

Problem Solving

Solve.

12. Hector has 20 rocket stickers. He wants to put 4 of them on each page of a scrapbook. How many pages will Hector need?

____

13. Brianna and 2 friends went to a star show at the planetarium. She paid $15 for all 3 tickets. How much did each ticket cost?

____

Name ______________________________

Divide by 2

Divide. Write a related multiplication fact.

1. $10 \div 2 =$ _____

____

2. $6 \div 2 =$ _____

____

3. $14 \div 2 =$ _____

____

4. $16 \div 2 =$ _____

____

5. $18 \div 2 =$ _____

____

6. $12 \div 2 =$ _____

____

7. $2\overline{)4} =$ _____

____

8. $2\overline{)16} =$ _____

____

Algebra Describe and complete the skip-counting patterns.

9. 16, 14, _____, 10

____

10. _____, 16, 14, 12

____

11. 36, 34, _____, 30

____

12. 158, 156, _____, 152

____

Problem Solving

Solve.

13. Franco records the tides when he visits his uncle who lives near the ocean. There are 2 high tides every day. Franco has recorded 16 high tides. How many days has Franco visited his uncle so far?

____

14. A book about astronauts costs $18. Gina and Kelly want to buy it together. How much money should each girl pay for the book if they each pay the same amount?

____

Name ____________________

Problem Solving: Skill

Choose an Operation

13-5 PRACTICE

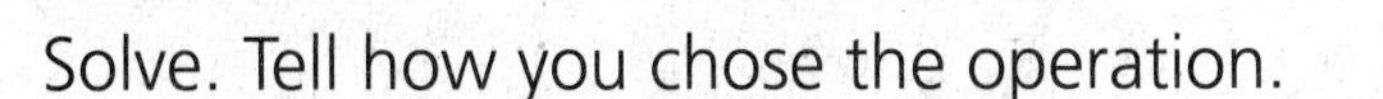

Solve. Tell how you chose the operation.

1. Paul is making a solar system notebook. Paul draws pictures of 9 planets. He draws 3 planets on each page. How many pages does he use?

2. Leroy has 15 pieces of clay. He will divide the clay equally to make models of 5 planets. How many pieces of clay will Leroy use to make each planet?

3. Naomi had 12 sun stickers. She gave 3 sun stickers to Jose. How many sun stickers does Naomi have left?

4. There are 45 children in the planetarium. They are sitting in rows of 9. How many rows of children are there?

Mixed Strategy Review

5. Brian draws 16 constellations. He gives away 4 drawings. How many drawings does Brian have left?

6. Janice uses 17 stars to draw the constellation Andromeda. She uses 8 stars to draw the constellation Cepheus. How many more stars did Janice use in drawing Andromeda than in Cepheus?

Name ______________________________

Divide by 5

14-1 PRACTICE

Divide.

1. $25 \div 5 =$ _____ **2.** $15 \div 5 =$ _____ **3.** $10 \div 5 =$ _____

4. $40 \div 5 =$ _____ **5.** $30 \div 5 =$ _____ **6.** $35 \div 5 =$ _____

7. $5 \div 5 =$ _____ **8.** $45 \div 5 =$ _____ **9.** $20 \div 5 =$ _____

10. $5\overline{)15}$ **11.** $5\overline{)30}$ **12.** $5\overline{)35}$ **13.** $5\overline{)25}$

14. $5\overline{)20}$ **15.** $5\overline{)5}$ **16.** $5\overline{)45}$ **17.** $5\overline{)40}$

Algebra Find each missing number.

18. $5 \times a = 20$

$a =$ _____

19. $30 \div d = 6$

$d =$ _____

20. $20 - n = 15$

$n =$ _____

21. $q \times 5 = 40$

$q =$ _____

22. $k - 5 = 40$

$k =$ _____

23. $20 \div p = 4$

$p =$ _____

24. $(7 + c) \div 5 = 3$

$c =$ _____

25. $(25 - 15) \div v = 2$

$v =$ _____

26. $(35 + s) \div 5 = 7$

$s =$ _____

Problem Solving

Solve.

27. Rudy spent $30 to buy 5 shuttle models. Each model costs the same amount. How much money did each model cost?

____

28. There are 40 people on the Space Rocket ride at the amusement park. Each car holds 5 people. All the cars are full. How many cars does the ride have?

____

Name ______________________________

Divide by 3

14-2 PRACTICE

Divide.

1. 18 ÷ 3 = _____ **2.** 9 ÷ 3 = _____ **3.** 6 ÷ 3 = _____

4. 24 ÷ 3 = _____ **5.** 3 ÷ 3 = _____ **6.** 21 ÷ 3 = _____

7. 12 ÷ 3 = _____ **8.** 27 ÷ 3 = _____ **9.** 15 ÷ 3 = _____

10. $3\overline{)12}$ **11.** $3\overline{)18}$ **12.** $3\overline{)6}$ **13.** $3\overline{)21}$

14. $3\overline{)27}$ **15.** $3\overline{)3}$ **16.** $3\overline{)15}$ **17.** $3\overline{)24}$

Algebra Complete.

18. Rule: Divide by 3

Input	Output
18	
24	
27	

19. Rule: Multiply by 3

Input	Output
7	
	27
1	

20. Rule: __________

Input	Output
14	11
12	9
18	15

Problem Solving

Solve.

21. Miss Gomez's 21 third-grade students work in 3 equal groups to make models of spacecraft. How many students are in each group?

____

22. Chuck and his 2 brothers read 15 books about the solar system. Each boy reads the same number of books. How many books did each boy read?

____

Name ____________________

Divide by 4

14-3 PRACTICE

Divide.

1. $12 \div 4 =$ _____ **2.** $8 \div 4 =$ _____ **3.** $20 \div 4 =$ _____

4. $28 \div 4 =$ _____ **5.** $24 \div 4 =$ _____ **6.** $4 \div 4 =$ _____

7. $36 \div 4 =$ _____ **8.** $32 \div 4 =$ _____ **9.** $16 \div 4 =$ _____

10. $4\overline{)16}$ **11.** $4\overline{)28}$ **12.** $4\overline{)4}$ **13.** $4\overline{)20}$ **14.** $4\overline{)40}$

15. $4\overline{)32}$ **16.** $4\overline{)8}$ **17.** $4\overline{)24}$ **18.** $4\overline{)36}$ **19.** $4\overline{)44}$

Algebra Complete.

20.

Rule: Multiply by 4	
Input	**Output**
6	
9	
7	

21.

Rule: Divide by 4	
Input	**Output**
32	
20	
	6

22.

Rule: _______	
Input	**Output**
16	4
28	7
36	9

Problem Solving

Solve. Use the data from the pictograph.

23. How many third-grade students went on the school trip?

____

24. There were 32 fourth-grade students on the school trip. How many symbols would you show on the graph for the fourth-grade students? Draw the symbols on the graph.

____

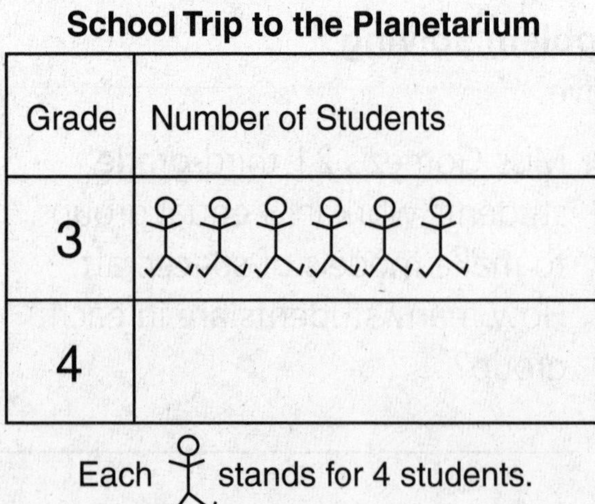

Name ______________________________

Divide with 0 and 1 • Algebra

Divide.

1. $0 \div 3 =$ ______ **2.** $5 \div 5 =$ ______ **3.** $4 \div 1 =$ ______

4. $9 \div 1 =$ ______ **5.** $3 \div 3 =$ ______ **6.** $5 \div 1 =$ ______

7. $8 \div 8 =$ ______ **8.** $0 \div 5 =$ ______ **9.** $0 \div 7 =$ ______

10. $5\overline{)0}$ **11.** $7\overline{)7}$ **12.** $4\overline{)0}$ **13.** $1\overline{)6}$ **14.** $2\overline{)0}$

15. $4\overline{)4}$ **16.** $1\overline{)4}$ **17.** $5\overline{)5}$ **18.** $3\overline{)0}$ **19.** $6\overline{)6}$

Algebra Write +, −, ×, or ÷ to make the number sentence true.

20. 7 ◯ 7 = 1 **21.** 9 ◯ 9 = 0 **22.** 6 ◯ 6 = 12

23. 5 ◯ 1 = 5 **24.** 0 ◯ 3 = 3 **25.** 4 ◯ 4 = 1

Problem Solving

Solve.

26. Jason buys 3 model rockets and shares them with 2 friends. How many rockets does each boy have?

____

27. Lisa has 3 empty key chains. How many keys does Lisa have?

____

28. Myra draws and cuts out 9 planets for a class project. She pastes each planet on a separate sheet of paper. How many sheets of paper did Myra use?

____

29. Alonzo has 1 key chain. It has 5 keys on it. How many keys does Alonzo have?

____

Name ______________________________

Problem Solving: Strategy
Act It Out

Solve.

1. Erica makes a solar system notebook. She draws 10 pictures that show the planets and the sun. Erica draws 2 pictures on each page. How many pages does Erica fill?

   ____

2. Maria and two other students build models of spaceships for a class project. Each student makes 5 spaceships. How many models did the students make in all?

   ____

3. A model of an Apollo command module costs $7. A model of a Gemini capsule costs $9. You have $20. If you buy both models, how much money will you have left?

   ____

4. There are 45 students in the third grade. Each group of 5 students is studying a different planet. How many planets is the third grade studying?

   ____

Mixed Strategy Review

5. **Science** Roger and a friend are studying the moons of Neptune. They each study the same number of moons. Neptune has 8 moons. How many moons does each friend study?

   ____

6. **Mental Math** The planetarium gives 4 shows on Saturday. The planetarium has 200 seats. How many people can see a planetarium show on Saturday?

   ____

7. **Write a problem** that you could solve by drawing a diagram or by writing a division sentence. Share it with others.

   ____

   ____

   ____

   ____

   ____

   ____

Name ____________________

Divide by 6 and 7

Divide.

1. $12 \div 6 =$ ______ **2.** $35 \div 7 =$ ______ **3.** $24 \div 6 =$ ______

4. $7 \div 7 =$ ______ **5.** $30 \div 6 =$ ______ **6.** $42 \div 7 =$ ______

7. $18 \div 6 =$ ______ **8.** $56 \div 7 =$ ______ **9.** $54 \div 6 =$ ______

10. $48 \div 6 =$ ______ **11.** $21 \div 7 =$ ______ **12.** $63 \div 7 =$ ______

13. $7\overline{)28}$ **14.** $6\overline{)36}$ **15.** $7\overline{)49}$ **16.** $6\overline{)24}$ **17.** $6\overline{)18}$

18. $6\overline{)48}$ **19.** $7\overline{)63}$ **20.** $7\overline{)21}$ **21.** $6\overline{)42}$ **22.** $7\overline{)14}$

23. $7\overline{)56}$ **24.** $7\overline{)42}$ **25.** $6\overline{)54}$ **26.** $6\overline{)30}$ **27.** $7\overline{)70}$

28. $6\overline{)12}$ **29.** $7\overline{)35}$ **30.** $6\overline{)6}$ **31.** $6\overline{)60}$ **32.** $7\overline{)7}$

Algebra Compare. Write $>$, $<$, or $=$.

33. $28 \div 7$ ◯ 5 **34.** $42 \div 7$ ◯ 5 **35.** $49 \div 7$ ◯ 8

36. $7 \div 7$ ◯ $6 \div 6$ **37.** $42 \div 6$ ◯ $42 \div 7$ **38.** $35 \div 7$ ◯ $30 \div 6$

39. $36 \div 6$ ◯ $35 \div 7$ **40.** $21 \div 7$ ◯ $18 \div 6$ **41.** $54 \div 6$ ◯ $56 \div 7$

Problem Solving

Solve.

42. Alberto plants 42 tree seedlings in 6 rows. Each row has the same number of tree seedlings. How many rows of tree seedlings does Alberto plant?

____

43. Six park rangers take 54 people on a tour of Great Bear National Park. Each ranger has the same number of tourists. How many people are in each group?

____

Name ________________________________

Problem Solving: Skill
Solve Multistep Problems

Solve.

1. There are 16 third graders and half as many fourth graders at the Nature Center. They are split into 4 equal groups. How many students are in each group?

   ____

2. During one afternoon at the Nature Center, 9 children watch a film. Twice as many children go on a hike. How many children are there in all?

   ____

3. Rob and his sister buy a "Save the Earth" game that costs $8. They also buy an animal calendar that costs half as much as the game. If they split the total cost evenly, how much does each pay?

   ____

4. Onida makes 6 endangered species greeting cards and sells the cards for $6 each. She splits the money equally among 4 environmental groups. How much money does each group get?

   ____

5. Carlos bundles 6 piles of newspapers each day. That is 3 times the number of piles that Lex bundles each day. How many piles do the two boys bundle in 4 days?

   ____

6. Jacob raises $7 by collecting cans and 3 times as much by collecting bottles. How much money does he raise in all?

   ____

7. Four friends make phone calls for the ecology club. Phone list A has 12 phone numbers. Phone list B has 4 more phone numbers than list A. If each friend makes an equal number of calls, how many phone calls does each friend make?

   ____

Name ____

Divide by 8 and 9

Divide.

1. $18 \div 9 =$ ______ **2.** $24 \div 8 =$ ______ **3.** $36 \div 9 =$ ______

4. $72 \div 8 =$ ______ **5.** $54 \div 9 =$ ______ **6.** $40 \div 8 =$ ______

7. $8 \div 8 =$ ______ **8.** $27 \div 9 =$ ______ **9.** $81 \div 9 =$ ______

10. $8\overline{)32}$ **11.** $9\overline{)9}$ **12.** $9\overline{)45}$ **13.** $8\overline{)16}$ **14.** $9\overline{)72}$

15. $9\overline{)63}$ **16.** $8\overline{)64}$ **17.** $9\overline{)54}$ **18.** $8\overline{)56}$ **19.** $8\overline{)48}$

Algebra Complete the tables.

20.

Rule: Divide by 9				
Input		72		81
Output	7		6	

21.

Rule: ______				
Input	40	48	56	72
Output	5	6	7	9

Problem Solving

Solve.

22. How many third-grade students volunteer for the Clean-Up Squad?

____

23. If 56 fourth-grade students volunteer, how many symbols should you show on the graph? Draw the symbols.

____

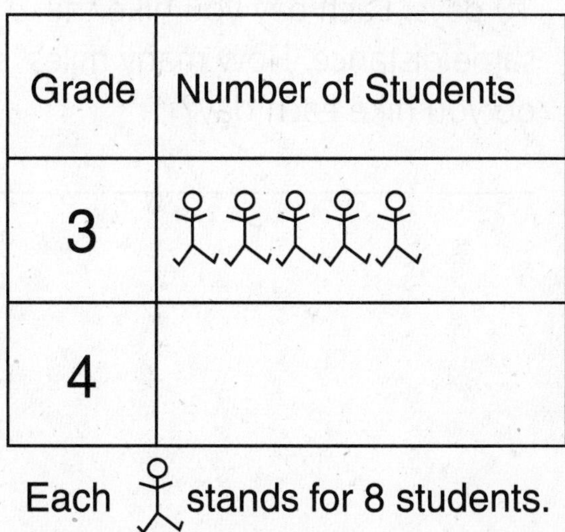

Volunteer Clean-Up Squad

Grade	Number of Students
3	(5 stick figures)
4	

Each (stick figure) stands for 8 students.

Name ________________________________

Explore Dividing by 10

Divide.

1. $70 \div 10 =$ ______ **2.** $10 \div 10 =$ ______ **3.** $60 \div 10 =$ ______

4. $20 \div 10 =$ ______ **5.** $30 \div 10 =$ ______ **6.** $90 \div 10 =$ ______

7. $50 \div 10 =$ ______ **8.** $80 \div 10 =$ ______ **9.** $40 \div 10 =$ ______

10. $10\overline{)20}$ **11.** $10\overline{)50}$ **12.** $10\overline{)10}$ **13.** $10\overline{)0}$ **14.** $10\overline{)30}$

15. $10\overline{)60}$ **16.** $10\overline{)90}$ **17.** $10\overline{)80}$ **18.** $10\overline{)40}$ **19.** $10\overline{)70}$

Problem Solving

Solve.

20. Thirty people paddle down the river on rafts. Each raft holds 10 people. How many rafts are on the river?

____

21. The Christo family spends $70 on 10 fishing permits. How much does each permit cost?

____

22. You hike a total of 60 miles in 10 days. Each day you hike the same distance. How many miles do you hike each day?

____

23. A group of park visitors spends $50 for 10 tickets for a raft ride. How much does each ticket cost?

____

Use with Grade 3, Chapter 15, Lesson 4, pages 334–335.

Name ____

Use a Multiplication Table

Complete the multiplication table. Use patterns to find the facts for 11 and 12.

×	0	1	2	3	4	5	6	7	8	9	10	11	12
0	0	0	0	0	0	0	0	0	0	0	0		
1	0	1	2	3	4	5	6	7	8	9	10		
2	0	2	4	6	8	10	12	14	16	18	20		
3	0	3	6	9	12	15	18	21	24	27	30		
4	0	4	8	12	16	20	24	28	32	36	40		
5	0	5	10	15	20	25	30	35	40	45	50		
6	0	6	12	18	24	30	36	42	48	54	60		
7	0	7	14	21	28	35	42	49	56	63	70		
8	0	8	16	24	32	40	48	56	64	72	80		
9	0	9	18	27	36	45	54	63	72	81	90		
10	0	10	20	30	40	50	60	70	80	90	100		
11													
12													

Find each missing number. Use the multiplication table.

1. $88 \div 11 = n$

$11 \times n = 88$

____

2. $108 \div 12 = b$

$12 \times b = 108$

____

3. $121 \div 11 = c$

$11 \times c = 121$

____

Divide.

4. $120 \div 10 =$ ________

5. $132 \div 12 =$ ________

6. $64 \div 8 =$ ________

7. $12\overline{)96}$

8. $11\overline{)99}$

9. $7\overline{)84}$

10. $12\overline{)144}$

11. $12\overline{)84}$

Problem Solving

Solve.

12. At the Wildflower Center, 12 students plant 144 bluebonnet seeds. Each student plants the same number of seeds. How many seeds does each student plant?

____

13. Volunteers at the Wildflower Center pick 108 flowers for bouquets. Each bouquet has 9 flowers. How many bouquets do they make?

____

Name ________________________________

Use Related Facts • Algebra

Write a fact family for each group of numbers.

1. 6, 7, 42

2. 7, 8, 56

3. 5, 9, 45

4. 6, 9, 54

5. 8, 9, 72

6. 9, 9, 81

Solve. Write a related fact.

7. $5 \times 6 =$ ______

8. $54 \div 6 =$ ______

9. $72 \div 8 =$ ______

10. $56 \div 7 =$ ______

11. $3 \times 9 =$ ______

12. $7 \times 5 =$ ______

Problem Solving

Solve.

13. Amanda takes a vacation for 8 weeks. How many days does she spend on vacation?

14. Jared and his brother hike 36 miles in 4 days. They hike the same distance each day. How many miles did they hike each day?

Name ____________________

Problem Solving: Strategy
Guess and Check

Use the guess-and-check strategy to solve.

1. A group of 14 scientists goes on a research trip. They sleep in 5 tents. Some are 2-person tents and some are 4-person tents. How many 2-person tents are there? How many 4-person tents?

____

____

2. **Language Arts** Kyle writes the spelling word *wilderness* over and over. When he is finished he has written the letters *e* and *s* a total of 28 times. How many times has he written the spelling word?

____

3. Two numbers have a product of 18 and quotient of 2. What are the numbers?

____

4. There are 8 animals in a field. Some of them are birds and some of them are bears. There are 22 legs. How many birds are there? How many bears are there?

____

Mixed Strategy Review

5. Ben buys binoculars with a hundred-dollar bill. With his change, he buys a canteen for $7. Now he has $43. What is the cost of the binoculars?

____

6. On Saturday, the recycling center receives 548 cans and 359 bottles. How many cans and bottles does it receive in all?

____

7. **Write a problem** that can be solved using the guess-and-check strategy. Share it with others.

____

____

____

____

____

____

Name ____________________

Explore the Mean

P 16-3 PRACTICE

Find the mean.

1. 1, 3 ______ **2.** 6, 4 ______ **3.** 4, 0, 5 ______

4. 1, 3, 2 ______ **5.** 4, 6, 5 ______ **6.** 0, 3, 6 ______

Find the mean. Use dots or connecting cubes to help.

7. 2, 2, 1, 3 ______ **8.** 0, 2, 3, 3 ______ **9.** 4, 6, 2, 4 ______

10. 2, 2, 4, 5, 2 ______ **11.** 8, 6 ______ **12.** 1, 2, 5, 4 ______

13. 3, 0, 5, 8 ______ **14.** 0, 8 ______ **15.** 5, 6, 7 ______

16. 5, 7, 0, 8 ______ **17.** 6, 0, 9 ______ **18.** 2, 1, 5, 4 ______

Show Your Work

Name ____________________

Find the Mean

1. number of roses in flower arrangements

4, 6, 3, 4, 3

____

2. number of tulips in flower arrangements

8, 6, 9, 9

____

3. prices of tickets

$3, $5, $4, $6, $2

____

4. number of cats adopted

5, 8, 7, 4, 3, 4, 4

____

5. $3, $5, $4, $8, $2, $6, $8, $4

____

6. 7, 11, 10, 9, 8, 10, 8

____

7. 14, 0, 16, 12, 8

____

8. 9, 11, 15, 13, 12,12

____

Solve. Use the data from the pictograph for problems 9–12.

	Number of Fish Caught
Marcus	🐟 🐟 🐟
Hailey	🐟 🐟
Lee	🐟 🐟 🐟 🐟
Sharon	🐟 🐟 🐟

🐟 = 2 fish

9. Who caught the greatest number of fish?

____

10. What is the mean number of fish caught?

____

11. What is the mode of the number of fish caught?

____

12. What is the range of the number of fish caught?

____

Name ____________________

Explore Multiplying Multiples of 10

Find the product.

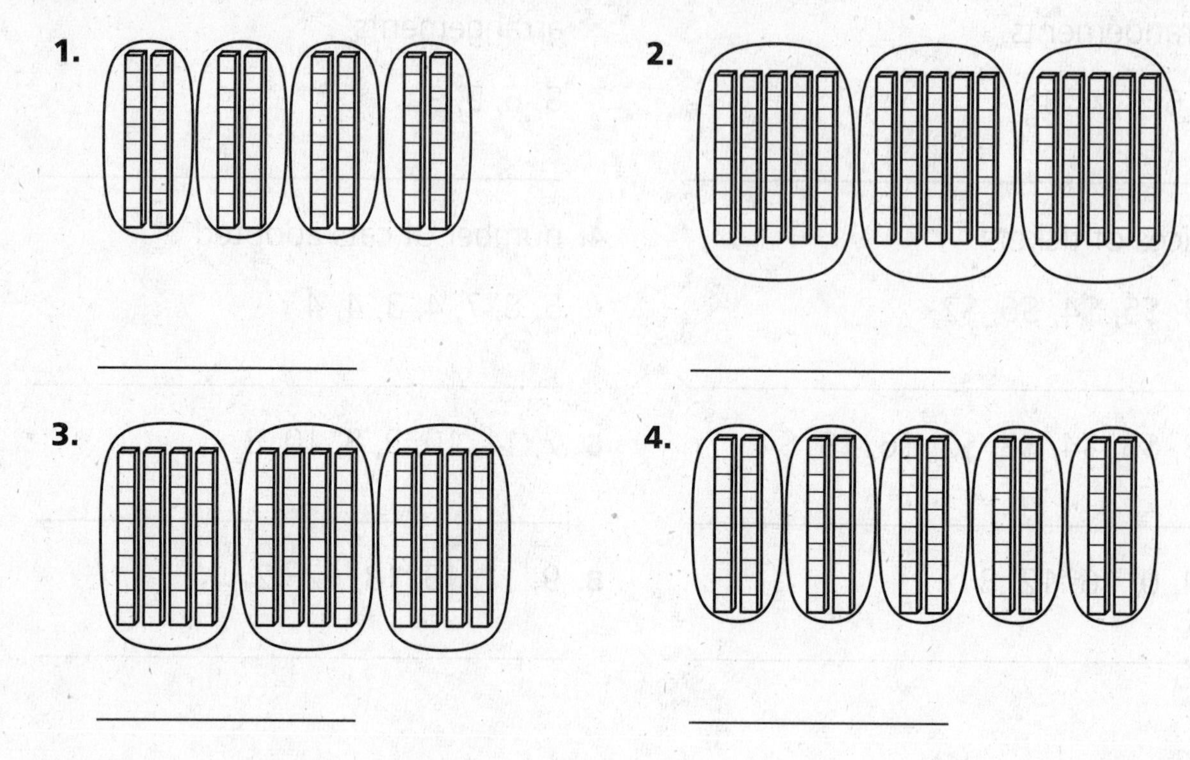

Multiply. You may use models.

5. $5 \times 30 =$ ____ 6. $5 \times 40 =$ ____ 7. $4 \times 70 =$ ____

8. $3 \times 80 =$ ____ 9. $6 \times 20 =$ ____ 10. $3 \times 60 =$ ____

11. $8 \times 40 =$ ____ 12. $9 \times 30 =$ ____ 13. $6 \times 30 =$ ____

14. $4 \times 30 =$ ____ 15. $8 \times 20 =$ ____ 16. $5 \times 60 =$ ____

17. $9 \times 20 =$ ____ 18. $2 \times 90 =$ ____ 19. $9 \times 40 =$ ____

Problem Solving

Solve.

20. Tara buys 3 rolls of stickers. There are 50 stickers on each roll. How many stickers does she buy?

____

21. Tara puts some of her stickers in a book. She fills 2 pages. Each page has 40 stickers on it. How many stickers are on those pages?

____

Name ______________________________

Multiplication Patterns • Algebra

Write the number that makes each sentence true.

1. $5 \times 2 =$ _____

   $5 \times$ _____ $= 100$

   $5 \times$ _____ $= 1{,}000$

   _____ $\times 2{,}000 = 10{,}000$

2. $3 \times 9 =$ _____

   $3 \times$ _____ $= 270$

   $3 \times 900 =$ _____

   $3 \times$ _____ $= 27{,}000$

3. $6 \times 7 =$ _____

   $6 \times$ _____ $= 420$

   $6 \times 700 =$ _____

   $6 \times$ _____ $= 42{,}000$

4. $3 \times$ _____ $= 15$

   $3 \times 50 =$ _____

   $3 \times$ _____ $= 1{,}500$

   $3 \times 5{,}000 =$ _____

5. $6 \times$ _____ $= 18$

   $6 \times 30 =$ _____

   $6 \times 300 =$ _____

   $6 \times$ _____ $= 18{,}000$

6. _____ $\times 3 = 24$

   _____ $\times 3 = 240$

   _____ $\times 3 = 2{,}400$

   _____ $\times 3 = 24{,}000$

Multiply. Use mental math.

7. $3 \times 40 =$ _______
8. $80 \times 5 =$ _______
9. $60 \times 5 =$ _______
10. $400 \times 6 =$ _______
11. $700 \times 6 =$ _______
12. $8 \times 7{,}000 =$ _______

Problem Solving

Solve.

13. A library spends $1,000 each month for new books. How much does it spend in 6 months?

    ____

14. Mrs. Thomas works in the library for 20 hours each week. How many hours does she work in 4 weeks?

    ____

15. Mr. Lin can put 30 books on one shelf. How many books can he put on 4 shelves?

    ____

16. Ellen designs bookmarks using her computer. She can print 20 bookmarks on one page. How many bookmarks can she print on 5 pages?

    ____

Name ______________________________

Estimate Products

Estimate each product. Tell how you estimate.

1. 344 × 4

2. $762 × 8

3. 681 × 7

4. 539 × 3

5. 133 × 9

6. 3,862 × 9

7. 3,245 × 5

8. $4,926 × 8

9. 2,540 × 3

10. 8,410 × 6

11. 2 × 249 ________

12. 7 × 74 ________

13. 4 × 508 ________

14. 5 × 632 ________

15. 3 × 628 ________

16. 5 × 457 ________

17. 8 × 5,298 ________

18. 2 × 1,328 ________

19. 6 × 278 ________

20. 8 × 99 ________

21. 2 × 4,829 ________

22. 4 × 86 ________

Algebra Estimate. Write > or < to make each sentence true.

23. 6 × 56 ◯ 420

24. 9 × 87 ◯ 720

25. 5 × 72 ◯ 300

Problem Solving

Solve.

26. Each day an average of 6,527 people come to the baseball card show. About how many people come to the show during the 7 days it is open?

____

27. If each person at the show spends $4 to get into the show, about how much is collected each day from 6,527 people?

____

Name ____

Problem Solving: Skill

Find an Estimate or Exact Answer

Solve. Tell if you found an estimate or exact answer.

1. Kate makes 9 bracelets. Each bracelet uses 20 shells. How many shells does she use?

   ____

2. John has 6 jars filled with shells. Each jar has 100 shells. How many shells does he have altogether?

   ____

3. Tracey buys 8 bags of shells. Each bag has 95 shells. Does she buy more than 800 shells?

   ____

4. Maria sells 8 paintings of shells for $28 each. How much money does she earn?

   ____

5. Kanesha fills an album with photos. The album has 8 pages. Each page holds 10 photos. How many photos are in the album?

   ____

6. Mike has $600 to spend on vacation. Does he have enough money to spend $79 each day for 7 days?

   ____

7. Sean works for 38 hours. He earns $6 per hour. How much money does he earn?

   ____

8. Terry stores her coin collection in jars. Each jar holds about 50 coins. If she has 5 filled jars, about how many coins does she have?

   ____

Mixed Strategy Review

9. Erin started her homework at 3:30. She watched television for 45 minutes and talked on the phone for 15 minutes. Her mom called her to dinner at 6:00. How much time did she spend on her homework?

   ____

10. Paul gives away 36 shells. He shares them equally among 4 friends. How many shells does each friend get?

   ____

Name ______________________________

Explore Multiplying 2-Digit Numbers

18-1 PRACTICE

Write a number sentence and then solve.

1.

2.

3.

4.

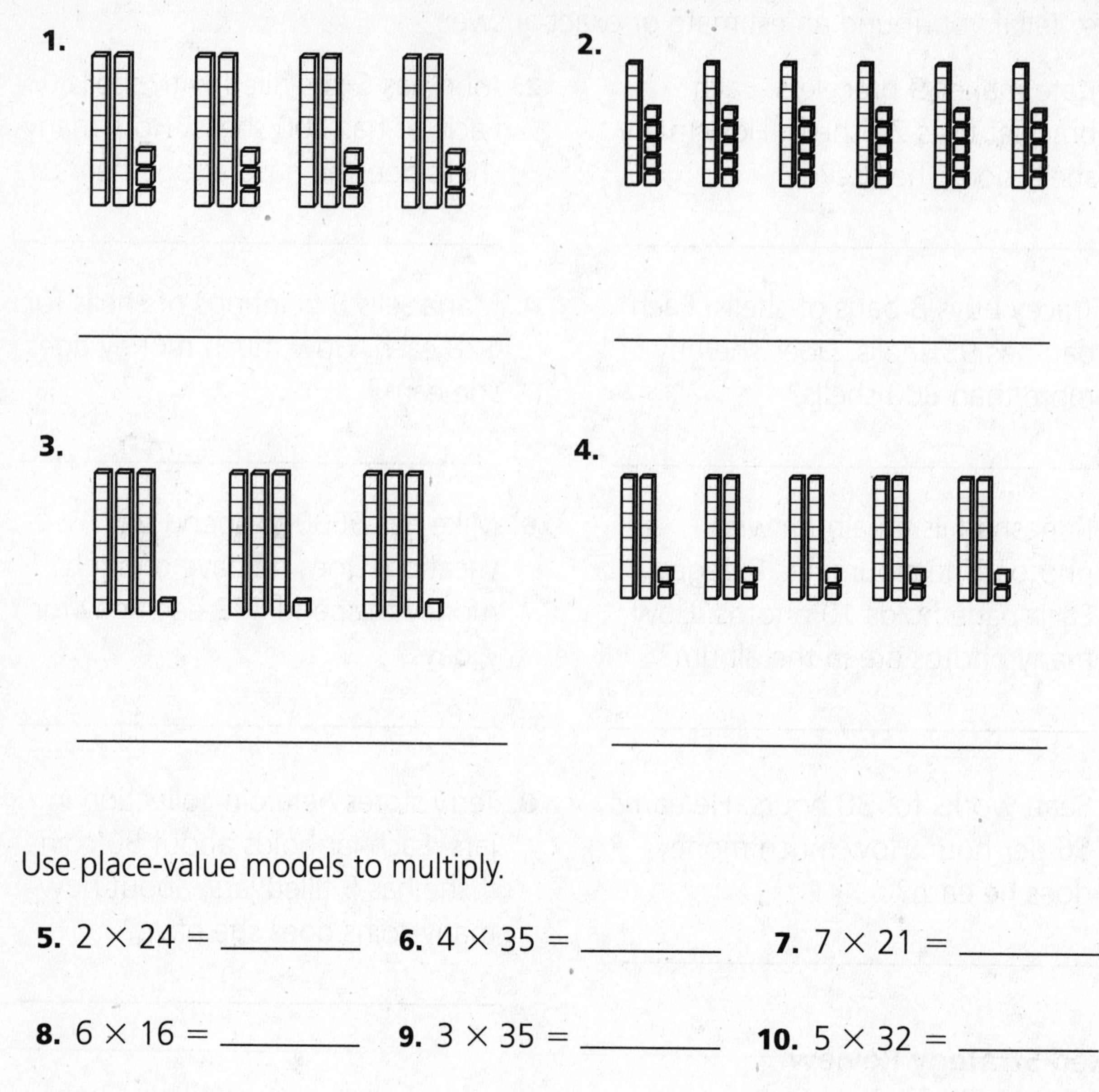

Use place-value models to multiply.

5. 2 × 24 = ________ **6.** 4 × 35 = ________ **7.** 7 × 21 = ________

8. 6 × 16 = ________ **9.** 3 × 35 = ________ **10.** 5 × 32 = ________

11. 3 × 43 = ________ **12.** 5 × 17 = ________ **13.** 6 × 33 = ________

14. 4 × 15 = ________ **15.** 8 × 22= ________ **16.** 3 × 42 = ________

Name ____________________

Multiply 2-Digit Numbers

Multiply.

1. $\begin{array}{r} 31 \\ \times\ 8 \\ \hline 248 \end{array}$

2. $\begin{array}{r} 38 \\ \times\ 5 \\ \hline \end{array}$

3. $\begin{array}{r} 28 \\ \times\ 2 \\ \hline \end{array}$

4. $\begin{array}{r} 43 \\ \times\ 7 \\ \hline \end{array}$

5. $\begin{array}{r} 17 \\ \times\ 8 \\ \hline \end{array}$

6. $\begin{array}{r} 24 \\ \times\ 8 \\ \hline \end{array}$

7. $\begin{array}{r} \$0.35 \\ \times\ 9 \\ \hline \end{array}$

8. $\begin{array}{r} 75 \\ \times\ 2 \\ \hline \end{array}$

9. $\begin{array}{r} 25 \\ \times\ 5 \\ \hline \end{array}$

10. $\begin{array}{r} 78 \\ \times\ 5 \\ \hline \end{array}$

11. $\begin{array}{r} 59 \\ \times\ 2 \\ \hline \end{array}$

12. $\begin{array}{r} 14 \\ \times\ 3 \\ \hline \end{array}$

13. $\begin{array}{r} 61 \\ \times\ 6 \\ \hline \end{array}$

14. $\begin{array}{r} 79 \\ \times\ 3 \\ \hline \end{array}$

15. $\begin{array}{r} \$0.44 \\ \times\ 9 \\ \hline \end{array}$

16. $\begin{array}{r} 18 \\ \times\ 5 \\ \hline \end{array}$

17. $\begin{array}{r} 64 \\ \times\ 2 \\ \hline \end{array}$

18. $\begin{array}{r} 36 \\ \times\ 7 \\ \hline \end{array}$

19. $6 \times \$92 =$ ________

20. $\$0.75 \times 9 =$ ________

21. $3 \times 85 =$ ________

22. $9 \times \$12 =$ ________

23. $2 \times 15 =$ ________

24. $32 \times 4 =$ ________

Problem Solving

Solve.

25. Becky charges $25 rent for each space at her flea market. If 8 people rent space, how much money does Becky get?

____

26. Mrs. Sands teaches 9 different classes at the high school. There are 36 students in each class. How many students does she teach?

____

Name ____

Problem Solving: Strategy

Make a Graph

On a separate piece of paper, make a pictograph for the data in the table. Include a key. Use data from the graph to solve problems 1 and 2.

Biggest Baseball Card Collections	
Name	**Number of Cards**
Amanda	475
Arnie	350
Bruce	525
Carol	600

1. Who collected the greatest number of cards?

____

2. Who collected the fewest number of cards?

____

3. Jack's coin collection includes 40 dimes, 35 nickels, and 75 pennies. Suppose you make a pictograph in which each symbol stands for 10 coins. How many symbols would you draw for each kind of coin?

____

____

____

4. Selma's coin collection includes 36 dimes, 78 nickels, and 102 pennies. Suppose you make a pictograph in which each symbol stands for 12 coins. How many symbols would you draw for each kind of coin?

____

____

Mixed Strategy Review

5. Two numbers have a product of 24 and a difference of 5. What are the numbers?

____

6. Mindy has twice as many teddy bears as Jolene. Together they have 75 teddy bears. How many teddy bears does each have?

____

____

Use with Grade 3, Chapter 18, Lesson 3, pages 388–389.

Name ____

Multiply Greater Numbers

Multiply.

1. $\begin{array}{r} 245 \\ \times\ 2 \\ \hline 490 \end{array}$

2. $\begin{array}{r} 121 \\ \times\ 6 \\ \hline \end{array}$

3. $\begin{array}{r} 240 \\ \times\ 7 \\ \hline \end{array}$

4. $\begin{array}{r} 324 \\ \times\ 4 \\ \hline \end{array}$

5. $\begin{array}{r} \$6.05 \\ \times\ 8 \\ \hline \end{array}$

6. $\begin{array}{r} 322 \\ \times\ 7 \\ \hline \end{array}$

7. $\begin{array}{r} 573 \\ \times\ 3 \\ \hline \end{array}$

8. $\begin{array}{r} 689 \\ \times\ 2 \\ \hline \end{array}$

9. $\begin{array}{r} \$4.95 \\ \times\ 5 \\ \hline \end{array}$

10 $\begin{array}{r} 225 \\ \times\ 9 \\ \hline \end{array}$

11. $\begin{array}{r} 304 \\ \times\ 8 \\ \hline \end{array}$

12. $\begin{array}{r} 923 \\ \times\ 4 \\ \hline \end{array}$

13. $\begin{array}{r} \$2,313 \\ \times\ 3 \\ \hline \end{array}$

14. $\begin{array}{r} 5,112 \\ \times\ 4 \\ \hline \end{array}$

15. $\begin{array}{r} 3,043 \\ \times\ 6 \\ \hline \end{array}$

16. $\begin{array}{r} \$10.45 \\ \times\ 8 \\ \hline \end{array}$

17. $\begin{array}{r} 1,623 \\ \times\ 7 \\ \hline \end{array}$

18. $\begin{array}{r} 2,418 \\ \times\ 9 \\ \hline \end{array}$

19. $\begin{array}{r} 9,372 \\ \times\ 5 \\ \hline \end{array}$

20. $\begin{array}{r} 2,094 \\ \times\ 9 \\ \hline \end{array}$

21. $2 \times 638 =$ ________

22. $6 \times 704 =$ ________

23. $2 \times \$2.25 =$ ________

24. $8 \times 1,976 =$ ________

25. $4 \times 2,430 =$ ________

26. $3 \times \$40.99 =$ ________

Problem Solving

Solve.

27. The field trip to the art museum costs $6.50. Mrs. King collects the money from the 6 students in her group. How much does she collect?

____

28. Each wing of the museum has 2,500 pieces of art on display. How many pieces of art are displayed in the 4 wings of the museum?

____

Name ____

Choose a Computation Method

P 18-5 PRACTICE

Multiply. Tell which method you use.

1. $5 \times 73 =$ ________ **2.** $7 \times 32 =$ ________ **3.** $8 \times 48 =$ ________

4. $4 \times 156 =$ ________ **5.** $3 \times 618 =$ ________ **6.** $9 \times 391 =$ ________

7. $604 \times 7 =$ ________ **8.** $208 \times 5 =$ ________ **9.** $109 \times 4 =$ ________

10. $4 \times 5{,}023 =$ ________ **11.** $1{,}307 \times 2 =$ ________ **12.** $9 \times 3{,}956 =$ ________

13. $2 \times 5{,}025 =$ ________ **14.** $4 \times 7{,}091 =$ ________ **15.** $7 \times 4{,}789 =$ ________

Algebra Compare. Write $<$ or $>$.

16. 4×28 ◯ 36×3

17. 5×73 ◯ 62×6

18. 3×95 ◯ 2×207

19. 8×350 ◯ 7×375

20. $3 \times 1{,}205$ ◯ $4 \times 1{,}025$

21. $8 \times 2{,}789$ ◯ $7 = 2{,}854$

Problem Solving

Solve.

22. Darla bought 3 paintings. Each painting cost $145. How much money did she spend?

____

23. There are 6 pictures on each page of Derrick's photo album. There are 72 pages in his album. How many pictures are there?

____

Name ____________________

Explore Dividing Multiples of 10

Write a division sentence. Then solve.

1. ____________ **2.** ____________

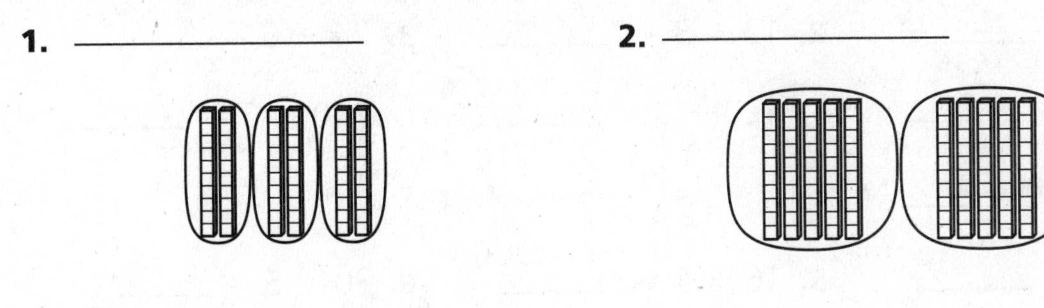

3. ____________ **4.** ____________

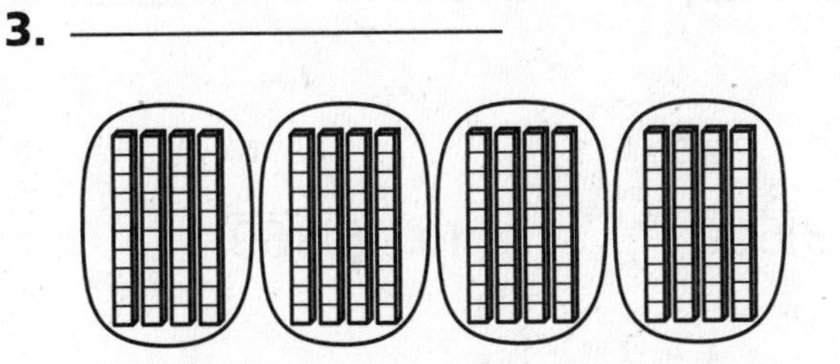

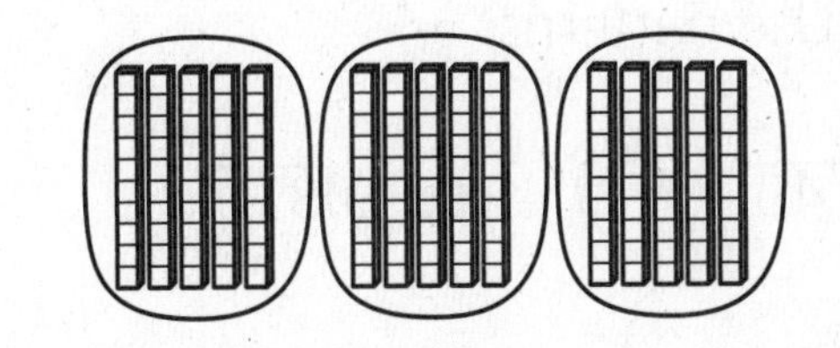

Divide. You may use place-value models.

5. $3\overline{)180}$ **6.** $4\overline{)240}$ **7.** $9\overline{)270}$ **8.** $2\overline{)140}$

9. $3\overline{)120}$ **10.** $8\overline{)640}$ **11.** $7\overline{)350}$ **12.** $4\overline{)360}$

13. $7\overline{)560}$ **14.** $3\overline{)240}$ **15.** $7\overline{)140}$ **16.** $6\overline{)240}$

17. $180 \div 2 =$ ______ **18.** $60 \div 2 =$ ______ **19.** $270 \div 3 =$ ______

20. $250 \div 5 =$ ______ **21.** $630 \div 7 =$ ______ **22.** $300 \div 5 =$ ______

23. $320 \div 4 =$ ______ **24.** $160 \div 2 =$ ______ **25.** $280 \div 4 =$ ______

Name ______________________________

Division Patterns • Algebra

19-2 PRACTICE

Write the number that makes each sentence true.

1. 8 ÷ 4 = 2
   80 ÷ 4 = 20
   800 ÷ 4 = 200

2. 12 ÷ 3 = ______
   120 ÷ 3 = ______

3. 24 ÷ 6 = ______
   240 ÷ 6 = ______

4. 63 ÷ 9 = ______
   630 ÷ 9 = ______

5. 16 ÷ 4 = ______
   160 ÷ 4 = ______

6. 30 ÷ 5 = ______
   300 ÷ 5 = ______

Divide. Use patterns.

7. 2)140
8. 9)810
9. 3)600
10. 6)360
11. 4)120
12. 9)90
13. 9)270
14. 5)250
15. 5)100
16. 4)200
17. 8)560
18. 9)180
19. 7)280
20. 6)480
21. 7)630
22. 8)400

23. 400 ÷ 5 = ______
24. 240 ÷ 4 = ______
25. 360 ÷ 4 = ______
26. 320 ÷ 8 = ______
27. 490 ÷ 7 = ______
28. 420 ÷ 6 = ______
29. 160 ÷ 8 = ______
30. 180 ÷ 6 = ______
31. 240 ÷ 3 = ______

Problem Solving

Solve.

32. An engineer estimates that a job will take 640 hours. If the engineer works 8 hours each day, how many days will it take to finish the job?

____

33. The engineer gives part of the job to 4 workers. If they each work the same number of hours and they work a total of 200 hours, how many hours do they each work?

____

Name ____

Estimate Quotients

Estimate.

1. $4\overline{)63}$ **2.** $7\overline{)580}$ **3.** $9\overline{)300}$ **4.** $8\overline{)390}$

5. $3\overline{)264}$ **6.** $3\overline{)130}$ **7.** $7\overline{)224}$ **8.** $5\overline{)392}$

9. $2\overline{)90}$ **10.** $2\overline{)900}$ **11.** $6\overline{)450}$ **12.** $6\overline{)370}$

13. $300 \div 4$ ________ **14.** $200 \div 3$ ________ **15.** $600 \div 7$ ________

16. $616 \div 8$ ________ **17.** $218 \div 5$ ________ **18.** $134 \div 2$ ________

Algebra Compare. Write $>$ or $<$.

19. $84 \div 2$ ◯ 40 **20.** $94 \div 3$ ◯ 40 **21.** $150 \div 3$ ◯ 60

22. $280 \div 90$ ◯ 30 **23.** $240 \div 5$ ◯ 60 **24.** $500 \div 8$ ◯ 70

Problem Solving

Solve.

25. There are 390 students going on a trip to a factory. They fill 5 buses. Each bus holds the same number of students. About how many students does each bus hold?

____

26. At the factory tour, the guide tells the students that each worker makes an average of 250 parts each day. About how many parts each hour does a worker make during an 8-hour day?

____

Name ________________

Explore Division

Use models to divide.

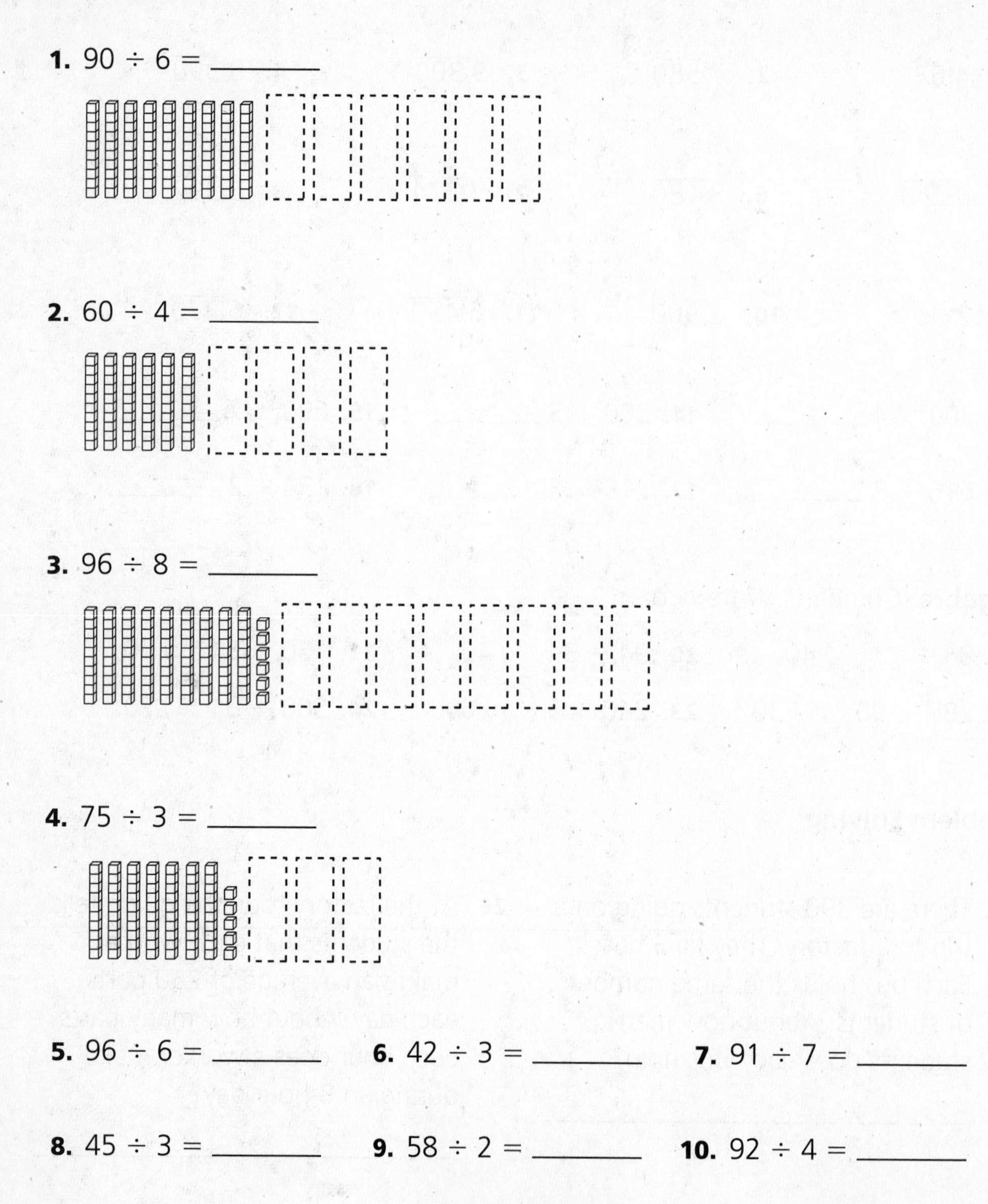

1. 90 ÷ 6 = ______

2. 60 ÷ 4 = ______

3. 96 ÷ 8 = ______

4. 75 ÷ 3 = ______

5. 96 ÷ 6 = ______

6. 42 ÷ 3 = ______

7. 91 ÷ 7 = ______

8. 45 ÷ 3 = ______

9. 58 ÷ 2 = ______

10. 92 ÷ 4 = ______

Name ____

Divide 2-Digit Numbers

Divide. Check your answer.

1. $5\overline{)64}$ 12 R4	**2.** $5\overline{)83}$	**3.** $4\overline{)65}$	**4.** $3\overline{)99}$
5. $8\overline{)94}$	**6.** $7\overline{)84}$	**7.** $2\overline{)86}$	**8.** $7\overline{)93}$
9. $6\overline{)93}$	**10.** $4\overline{)77}$	**11.** $4\overline{)96}$	**12.** $8\overline{)98}$
13. $3\overline{)95}$	**14.** $6\overline{)87}$	**15.** $3\overline{)85}$	**16.** $2\overline{)\$62}$
17. $6\overline{)78}$	**18.** $2\overline{)\$92}$	**19.** $5\overline{)81}$	**20.** $2\overline{)\$30}$

21. $97 \div 7 =$ ______   **22.** $49 \div 3 =$ ______   **23.** $51 \div 2 =$ ______

24. $88 \div 5 =$ ______   **25.** $99 \div 8 =$ ______   **26.** $92 \div 7 =$ ______

27. $70 \div 4 =$ ______   **28.** $95 \div 8 =$ ______   **29.** $\$72 \div 4 =$ ______

30. $61 \div 2 =$ ______   **31.** $64 \div 3 =$ ______   **32.** $73 \div 6 =$ ______

Problem Solving

Solve.

33. Dave earns $70 for cleaning a house. He splits the money equally with his helper. How much money do they each get?

____

34. Ruby earns $96 in tips at her job as a waitress. She divides her money into 3 equal amounts for taxes, spending, and saving. How much money does she have for spending?

____

Name ____

Problem Solving: Skill
Interpret the Remainder

Solve.

1. The park volunteers collect $33 to buy plants. Each plant costs $6. There is no sales tax. If the volunteers buy as many plants as they can, how much money will be left?

   ____

2. Four plants can fit in a window box. Janet buys 25 plants. How many window boxes does she need? How many plants are left?

   ____

3. The fire department expects 31 people at its picnic. A cake serves 8 people. How many cakes does the fire department need for the picnic?

   ____

4. Tickets to the Concert Under the Stars cost $9 each. Susie has $40. If she buys as many tickets as she can, how much money will she have left?

   ____

5. Each picnic table can seat 6 people. How many tables are needed for 31 people?

   ____

6. There are 43 volunteers. They travel in cars that hold 5 people each. How many cars do they need?

   ____

Mixed Strategy Review

Solve.

7. Hot dog buns come in packages of 8. The fire department buys 50 packages of buns for the picnic. How many hot dogs are needed?

   ____

8. There are 8 seats in each of the 23 rows of seats at the town meeting. How many people can the meeting hold?

   ____

Name ____________________

Divide 3-Digit Numbers

Divide. Check your answer.

1. 4)456 (114) **2.** 7)389 **3.** 6)297 **4.** 6)$5.28

5. 3)$6.81 **6.** 5)290 **7.** 2)578 **8.** 3)263

9. 5)745 **10.** 4)$8.68 **11.** 9)253 **12.** 7)$4.06

13. 4)368 **14.** 9)294 **15.** 3)573 **16.** 6)188

17. 7)$5.04 **18.** 8)576 **19.** 2)924 **20.** 6)837

21. 875 ÷ 7 = ______ **22.** 528 ÷ 3 = ______ **23.** $3.85 ÷ 5 = ______

24. 974 ÷ 2 = ______ **25.** $8.52 ÷ 3 = ______ **26.** 429 ÷ 8 = ______

27. 637 ÷ 8 = ______ **28.** 204 ÷ 3 = ______ **29.** 422 ÷ 5 = ______

30. $8.96 ÷ 2 = ______ **31.** 476 ÷ 4 = ______ **32.** 234 ÷ 2 = ______

Problem Solving

Solve.

33. Betsy and her brother split the cost of lemons for their lemonade stand. They spent a total of $2.98. How much did each pay?

____

34. The lemonade container holds 128 ounces. How many 8-ounce glasses of lemonade does the container hold?

____

Name ______________________________

Quotients with Zeros

20-2 PRACTICE

Divide. Check your answer.

60 R5

1. 6)365	**2.** 4)83	**3.** 4)360	**4.** 3)241
5. 6)620	**6.** 2)416	**7.** 4)804	**8.** 5)154
9. 3)922	**10.** 5)303	**11.** 4)423	**12.** 2)$6.10
13. 7)633	**14.** 8)647	**15.** 5)$5.40	**16.** 3)923
17. 2)810	**18.** 9)280	**19.** 8)323	**20.** 7)724

21. 313 ÷ 3 = ________ **22.** 152 ÷ 5 = ________ **23.** 720 ÷ 7 = ________

24. 722 ÷ 9 = ________ **25.** 606 ÷ 3 = ________ **26.** 325 ÷ 8 = ________

27. 121 ÷ 2 = ________ **28.** $8.08 ÷ 8 = ________ **29.** $9.36 ÷ 9 = ________

30. $4.06 ÷ 2 = ________ **31.** 648 ÷ 6 = ________ **32.** 542 ÷ 6 = ________

Problem Solving

Solve.

33. A fundraiser raised $816 for 4 charities. If each charity gets the same amount of money, how much do they each get?

____

34. Jane spent $5.25 at the fundraiser. She spent the same amount of money at each of 5 booths. How much did she spend at each booth?

____

Name ____________________

Choose a Computation Method

Divide. Tell which method you use.

1. $432 \div 4 =$ ______
2. $376 \div 6 =$ ______
3. $298 \div 8 =$ ______
4. $308 \div 2 =$ ______
5. $527 \div 7 =$ ______
6. $483 \div 3 =$ ______
7. $918 \div 4 =$ ______
8. $409 \div 6 =$ ______
9. $722 \div 3 =$ ______
10. $805 \div 5 =$ ______
11. $716 \div 4 =$ ______
12. $256 \div 8 =$ ______
13. $8\overline{)682}$
14. $3\overline{)128}$
15. $6\overline{)543}$

Algebra Find each missing number.

16. $804 \div k = 134$ ______
17. $n \div 8 = 72$ ______
18. $284 \div k = 71$ ______
19. $882 \div x = 126$ ______
20. $y \div 7 = 89$ ______
21. $416 \div k = 52$ ______

Problem Solving

Solve.

22. The art club shows 138 paintings at its annual exhibit. Each member contributes the same number of paintings. Each member contributes 6 paintings. How many members are in the club?

______

23. The art club displays 104 paintings on 4 walls. The same number of paintings hang on each wall. How many paintings are displayed on each wall?

______

Name ____________________

Problem Solving: Strategy

Choose a Strategy

Choose a strategy to solve the problem.

1. On Peapack's Park Day, volunteers work in the park. One volunteer has 8 boxes of plants. There are 12 plants in each box. If the volunteer puts the plants in rows of 6, how many rows can he make?

   ____

2. Peapack's town square is surrounded by 64 trees. The same number of trees are on each of the 4 sides. The trees on each side are divided into 2 equal rows. How many trees are in each row?

   ____

3. Jenny takes a photo of the town square. She makes a square frame for the photo. Each side of the frame is 12 inches long. How many inches around is the frame?

   ____

4. Some volunteers are building picnic tables. The tables can seat 4 adults or 6 children. How many adults can sit at 6 tables? How many children?

   ____

Mixed Strategy Review

5. This year, a town sells tickets to the picnic to 252 adults and 518 children. Last year, there were 695 people at the picnic. How many more people are there this year than last year?

   ____

6. **Art** The third grade class makes a mural for the train station. The mural is 30 feet long. The mural is divided into 6 equal sections. How many feet long is each section?

   ____

7. **Write a problem** that you could solve by drawing a diagram or by writing a division sentence. Share it with others.

   ____

   ____

   ____

   ____

Name ____________________

Explore Lengths

Use a nonstandard unit and a ruler to measure. Measure to the nearest inch. Write the length.

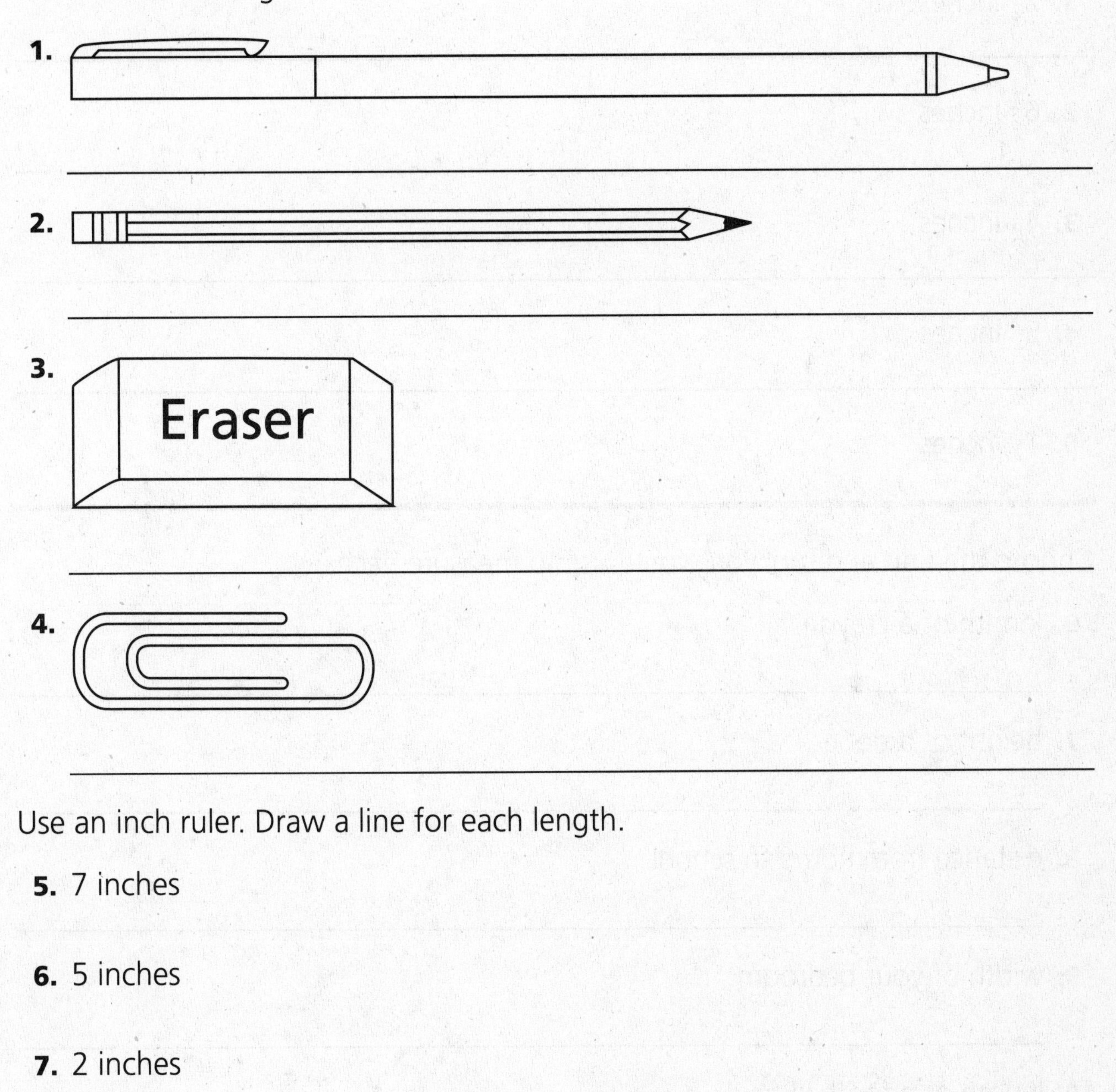

Use an inch ruler. Draw a line for each length.

5. 7 inches

6. 5 inches

7. 2 inches

8. 1 inch

9. 3 inches

10. 4 inches

Name ______________________________

Explore Customary Units of Length

Use an inch ruler. Draw a line for each length.

1. $2\frac{1}{2}$ inches

____

2. $6\frac{1}{2}$ inches

____

3. $4\frac{1}{2}$ inches

____

4. $5\frac{1}{2}$ inches

____

5. $1\frac{1}{2}$ inches

Choose the unit and tool you would use to measure each.

6. length of a crayon

____

7. height of a door

____

8. distance from home to school

____

9. width of your bedroom

____

10. length of a football field

____

11. thickness of a book

____

Name ____________________

Customary Units of Capacity

Ring the letter of the better estimate.

1.

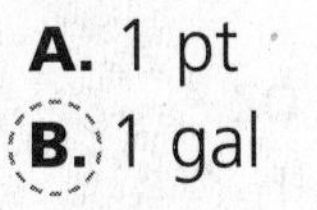

A. 1 pt
B. 1 gal

2.

A. 1 c
B. 1 qt

3.

A. 1 c
B. 1 pt

4.

A. 5 pt
B. 5 gal

5.

A. 2 c
B. 2 qt

6.

A. 200 qt
B. 200 gal

Problem Solving

Solve.

7. Jonathan drank 6 cups of water one day. Did he drink more or less than a quart of water?

____

8. Louisa filled a pot with water to make soup for the family. Did she put in 1 cup or 1 gallon of water?

____

Name ____________________

Customary Units of Weight

P 21-4 PRACTICE

Ring the letter of the better estimate.

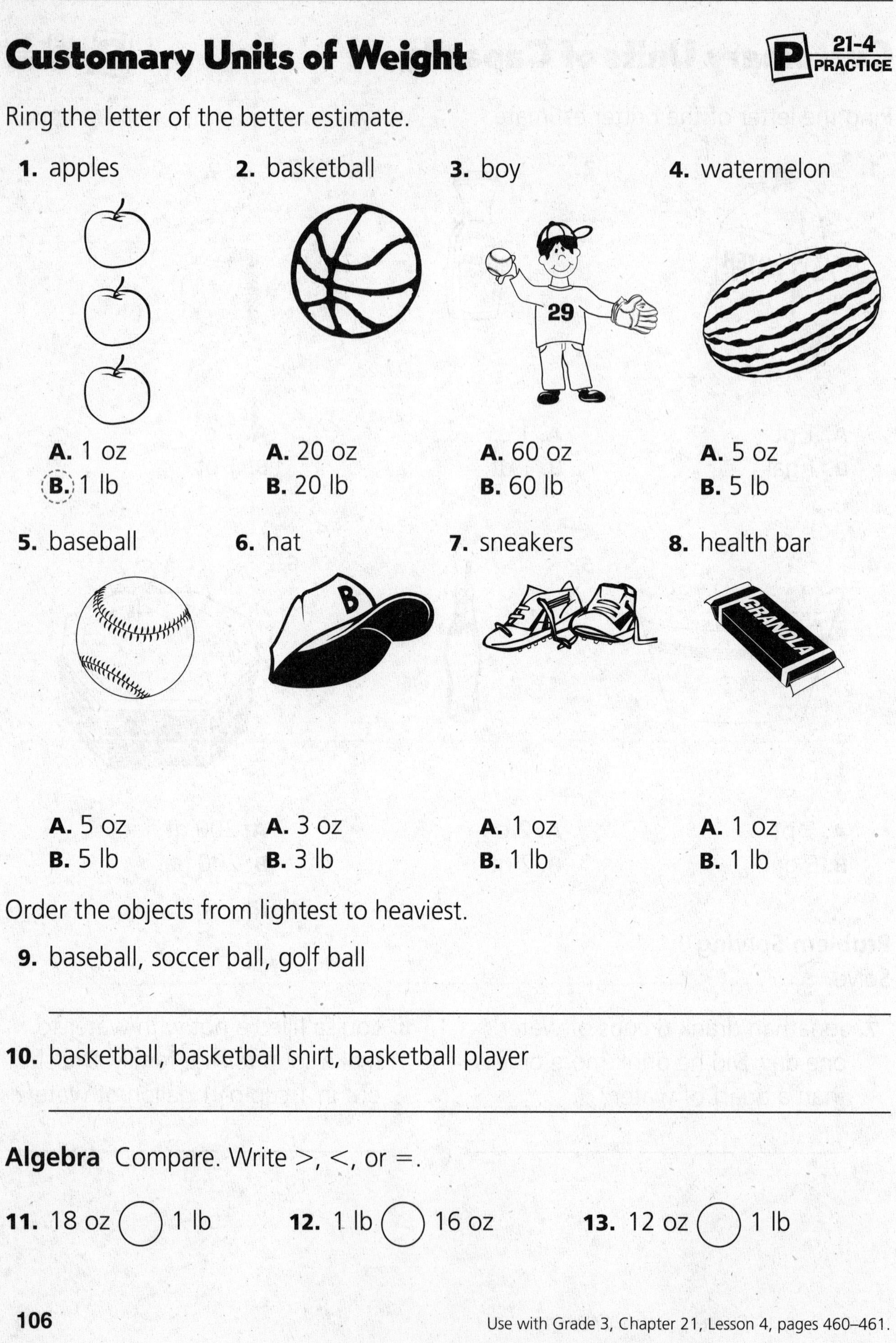

1. apples

A. 1 oz
B. 1 lb

2. basketball

A. 20 oz
B. 20 lb

3. boy

A. 60 oz
B. 60 lb

4. watermelon

A. 5 oz
B. 5 lb

5. baseball

A. 5 oz
B. 5 lb

6. hat

A. 3 oz
B. 3 lb

7. sneakers

A. 1 oz
B. 1 lb

8. health bar

A. 1 oz
B. 1 lb

Order the objects from lightest to heaviest.

9. baseball, soccer ball, golf ball

____

10. basketball, basketball shirt, basketball player

____

Algebra Compare. Write >, <, or =.

11. 18 oz ◯ 1 lb

12. 1 lb ◯ 16 oz

13. 12 oz ◯ 1 lb

Name ______________________________

Convert Customary Units

P 21-5 PRACTICE

Write the number that makes each sentence true.

1. 2 ft = _____ in. **2.** 2 yd = _____ ft **3.** 2 c = _____ pt

4. 5 lb = _____ oz **5.** 4 qt = _____ pt **6.** 6 c = _____ pt

7. 12 ft = _____ yd **8.** 16 pt = _____ qt **9.** 8 qt = _____ gal

10. _____ pt = 28 c **11.** _____ oz = 2 lb **12.** _____ gal = 12 qt

13. _____ qt = 4 pt **14.** _____ qt = 4 gal **15.** _____ c = 18 pt

16. _____ ft = 3 yd **17.** _____ qt = 9 gal **18.** _____ in. = 4 ft

Algebra Compare. Write >, <, or =.

19. 1 gal ○ 2 pt **20.** 1 ft ○ 1 yd **21.** 1 lb ○ 8 oz

22. 6 c ○ 2 pt **23.** 2 qt ○ 4 pt **24.** 36 in. ○ 2 ft

25. 2 gal ○ 10 qt **26.** 12 in. ○ 1 yd **27.** 6 pt ○ 3 qt

28. 18 in. ○ 1 ft **29.** 2 lb ○ 24 oz **30.** 20 c ○ 12 pt

Problem Solving

Solve.

31. In baseball, the pitcher's mound is 6 inches more than 20 yd from home plate. How many feet is this?

____

32. In baseball, the distance between first and second base is 90 ft. How many yards is this?

____

Name ____________________

Problem Solving: Skill
Check for Reasonableness

Solve. Explain your answer.

1. Ned shoots a basket from 10 feet. Demi shoots a basket from 114 inches. Ned says that his shot was from a greater distance than Demi's shot. Is this a reasonable statement?

____

____

2. David runs 440 yards. John runs 1,320 feet. John says he ran farther than David. Is his statement reasonable?

____

____

3. Jason throws a ball 15 yards. Virginia throws it 45 feet. Virginia says that her throw is just as long as Jason's. Is her statement reasonable?

____

____

4. José throws a shot-put ball 10 feet 9 inches. Marcus throws a shot-put ball 129 inches. Marcus says that he threw the ball farther. Is his statement reasonable?

____

____

5. A lap around the field is 800 yards. Paula runs 1 lap and says she has run more than 2,000 feet. Is her statement reasonable?

____

____

6. The track is 400 yards long. Donna says that this is less than 1,000 feet. Is her statement reasonable?

____

____

Mixed Strategy Review

7. Nick buys 2 gallons of orange juice for $9.20. How much does 1 quart of orange juice cost?

____

8. Bonnie high-jumps 28 inches. Selma jumps 3 feet. How much higher did Selma high-jump than Bonnie?

____

Name ____________________

Explore Metric Units of Length

P 22-1 PRACTICE

Estimate. Then measure to the nearest centimeter.

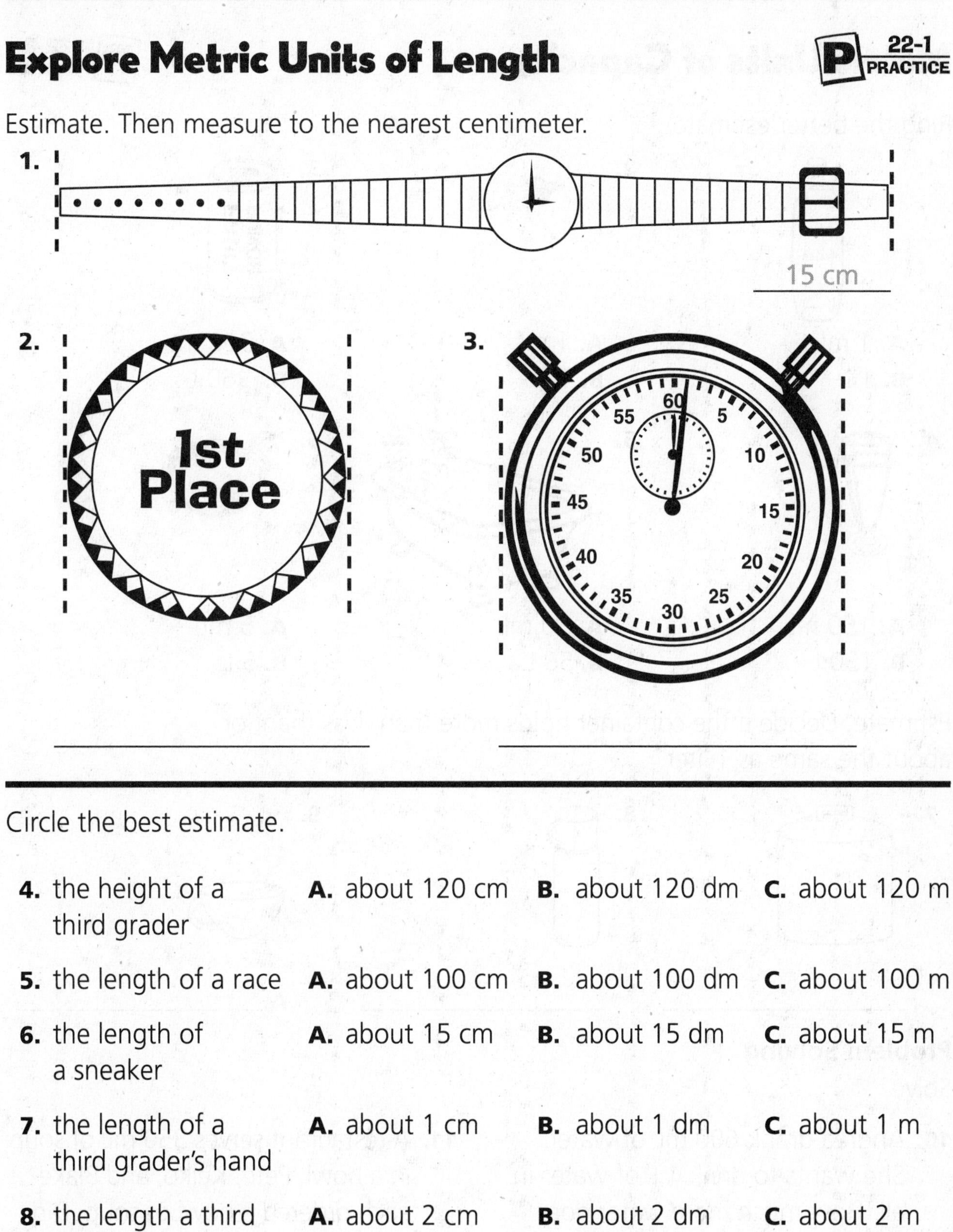

1. 15 cm

2. ____________

3. ____________

Circle the best estimate.

4. the height of a third grader — **A.** about 120 cm **B.** about 120 dm **C.** about 120 m

5. the length of a race — **A.** about 100 cm **B.** about 100 dm **C.** about 100 m

6. the length of a sneaker — **A.** about 15 cm **B.** about 15 dm **C.** about 15 m

7. the length of a third grader's hand — **A.** about 1 cm **B.** about 1 dm **C.** about 1 m

8. the length a third grader can jump — **A.** about 2 cm **B.** about 2 dm **C.** about 2 m

9. the length of a jump rope — **A.** about 5 cm **B.** about 5 dm **C.** about 5 m

Name ____________________

Metric Units of Capacity

Ring the better estimate.

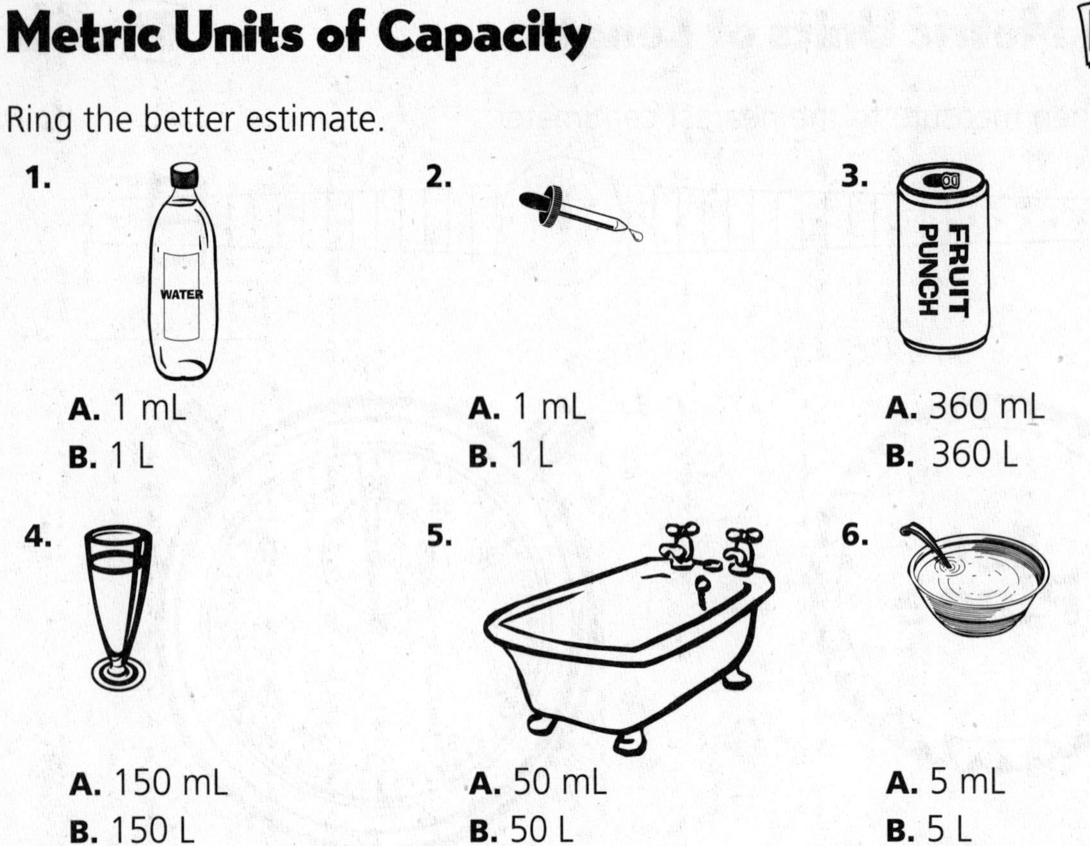

1. **A.** 1 mL
   B. 1 L

2. **A.** 1 mL
   B. 1 L

3. **A.** 360 mL
   B. 360 L

4. **A.** 150 mL
   B. 150 L

5. **A.** 50 mL
   B. 50 L

6. **A.** 5 mL
   B. 5 L

Estimate. Decide if the container holds more than, less than, or about the same as 1 liter.

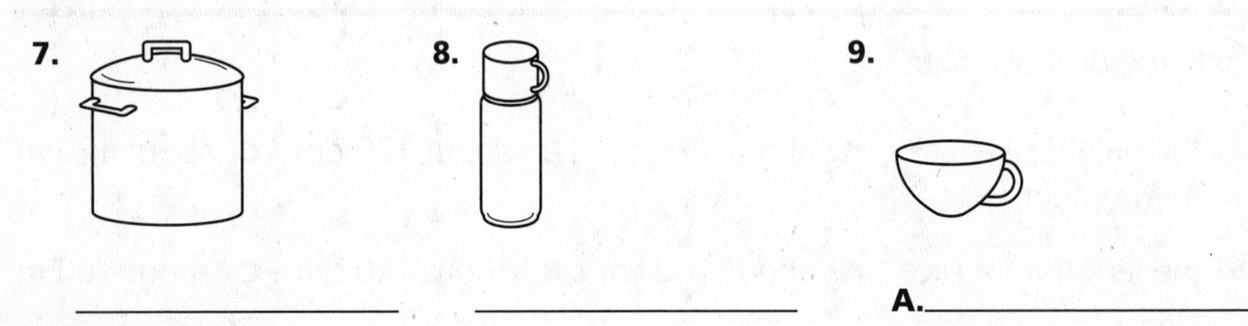

7. ____________________

8. ____________________

9. **A.** ____________________

Problem Solving

Solve.

10. Andrea drank 600 mL of water. She wants to drink 1 L of water in all. How much more water does she need to drink?

    ____

11. A restaurant serves 350 mL of soup in a bowl. Pete, Keiko, and Blake each ordered a bowl of soup. Did they order more than, less than, or 1 liter of soup? Explain.

    ____

Use with Grade 3, Chapter 22, Lesson 2, pages 474–475.

Name ______________________________

Metric Units of Mass

P 22-3 PRACTICE

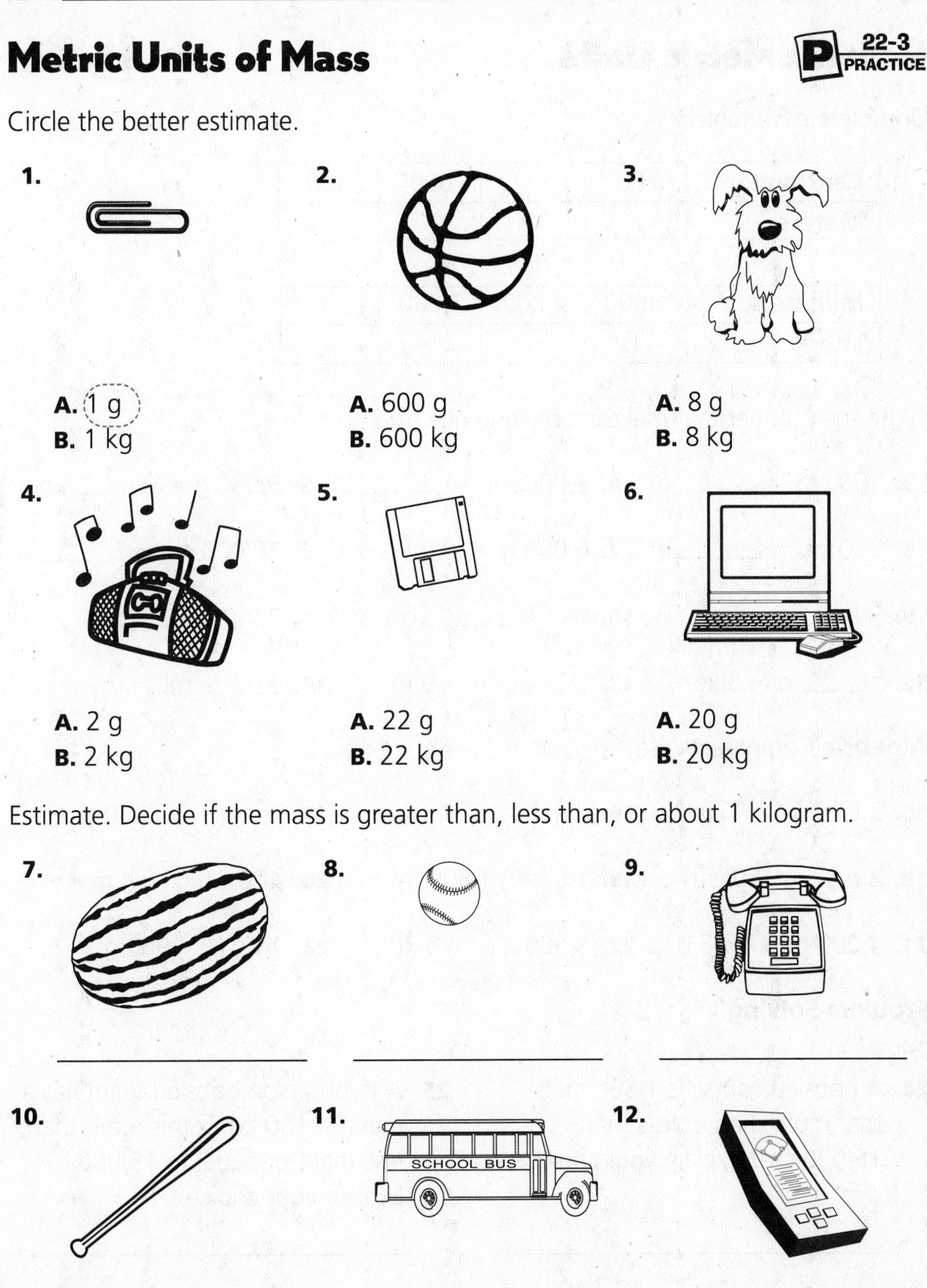

Circle the better estimate.

1.
A. 1 g
B. 1 kg

2.
A. 600 g
B. 600 kg

3.
A. 8 g
B. 8 kg

4.
A. 2 g
B. 2 kg

5.
A. 22 g
B. 22 kg

6.
A. 20 g
B. 20 kg

Estimate. Decide if the mass is greater than, less than, or about 1 kilogram.

7. ______________

8. ______________

9. ______________

10. ______________

11. ______________

12. ______________

Name ____

Convert Metric Units

Complete each table.

1.

Centimeters	200		600	800
Meters	2	4		

2.

Milliliters	1,000	2,000	3,000	
Liters	1			4

Write the number that makes each sentence true.

3. 100 cm = ______ m **4.** 2,000 mL = ______ L **5.** 3,000 g = ______ kg

6. 500 cm = ______ m **7.** 6,000 g = ______ kg **8.** 10,000 mL = ______ L

9. 3 L = ______ mL **10.** 6 m = ______ cm **11.** 2 kg = ______ g

12. ______ g = 5 kg **13.** ______ cm = 9 m **14.** ______ mL = 7 L

Algebra Compare. Write >, <, or =.

15. 2 L ◯ 1,000 mL **16.** 3 m ◯ 250 cm **17.** 2 kg ◯ 3,000 g

18. 2 m ◯ 180 cm **19.** 1 L ◯ 1,500 mL **20.** 400 cm ◯ 4 m

21. 4,000 mL ◯ 5 L **22.** 5,500 g ◯ 5 kg **23.** 1 kg ◯ 900 g

Problem Solving

Solve.

24. A bobsled course is 1,500 m. Is that more than or less than 15,000 cm? Explain your choice.

____

____

25. A two-person bobsled team has a mass of 150 kg. Is this more than, less than, or equal to 15,000 g? Explain your choice.

____

____

Use with Grade 3, Chapter 22, Lesson 4, pages 478–480.

Name ____________________

Problem Solving: Strategy

Logical Reasoning

Use logical reasoning to solve.

1. Coach Jack has a 5-liter bottle and a 4-liter bottle. How can he use the bottles to get exactly 11 liters of water in a cooler?

____

____

____

____

2. Coach Mary has a 5-gallon bottle and a 3-gallon bottle. How can she use the bottles to get exactly 14 gallons of water in a cooler?

____

____

____

____

3. Dan, Michael, and Jerry play different sports. One plays tennis, another plays baseball, and the third is on the swimming team. Michael and Jerry play sports that use balls. Michael does not play baseball. Who plays baseball?

____

4. Don, Shari, Steve, and Ellen are in line for football tickets. The first person in line is a girl. Shari is ahead of Steve, but not ahead of Don. List the names in order from first to last in line.

____

Mixed Strategy Review

5. A small van has 4 rows of seats. Each row can seat 3 people. How many people in all can 2 vans hold?

____

6. Leah is 5 feet tall. Her brother Jamie is 50 inches tall. How much taller is Leah than Jamie?

____

Name ______________________________

Temperature

Use the thermometer at the right to complete each exercise.

Choose the more reasonable temperature.

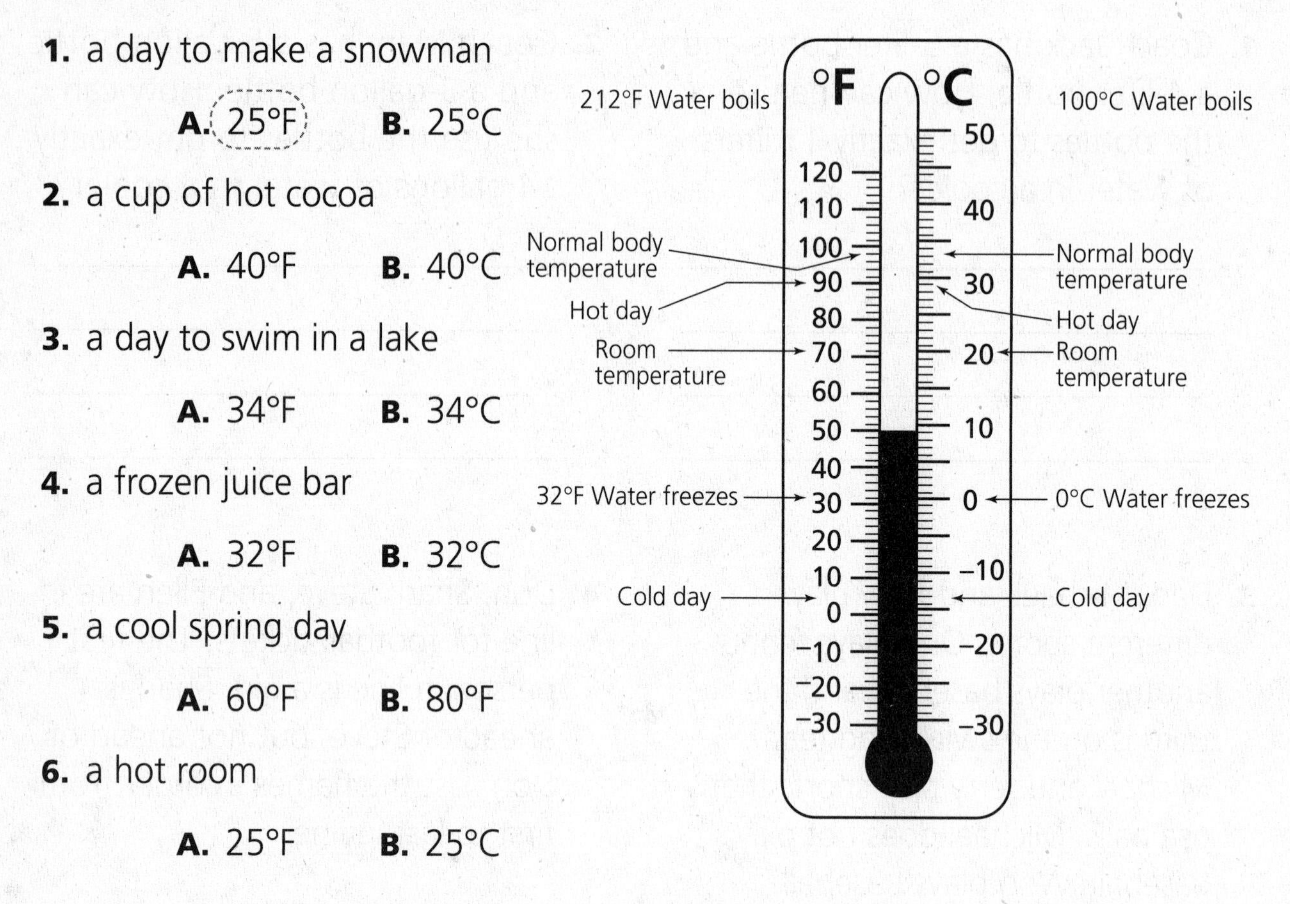

1. a day to make a snowman

A. 25°F **B.** 25°C

2. a cup of hot cocoa

A. 40°F **B.** 40°C

3. a day to swim in a lake

A. 34°F **B.** 34°C

4. a frozen juice bar

A. 32°F **B.** 32°C

5. a cool spring day

A. 60°F **B.** 80°F

6. a hot room

A. 25°F **B.** 25°C

Decide which kind of clothing you should wear.
Write T-shirt, sweatshirt, or heavy jacket.

7. 30°F ____________ **8.** 30°C ____________ **9.** 5°C ____________ **10.** 50°F ____________

Problem Solving

Solve.

11. Michelle wants to ice skate in the park. Is the temperature above or below 30°F?

____

12. Alex wants to go to the beach. The temperature outside is 25°C. Will Alex be able to go swimming? Explain your answer.

____

Name ______________________________

3-Dimensional Figures

P 23-1 PRACTICE

Name the 3-dimensional figure each object looks like.

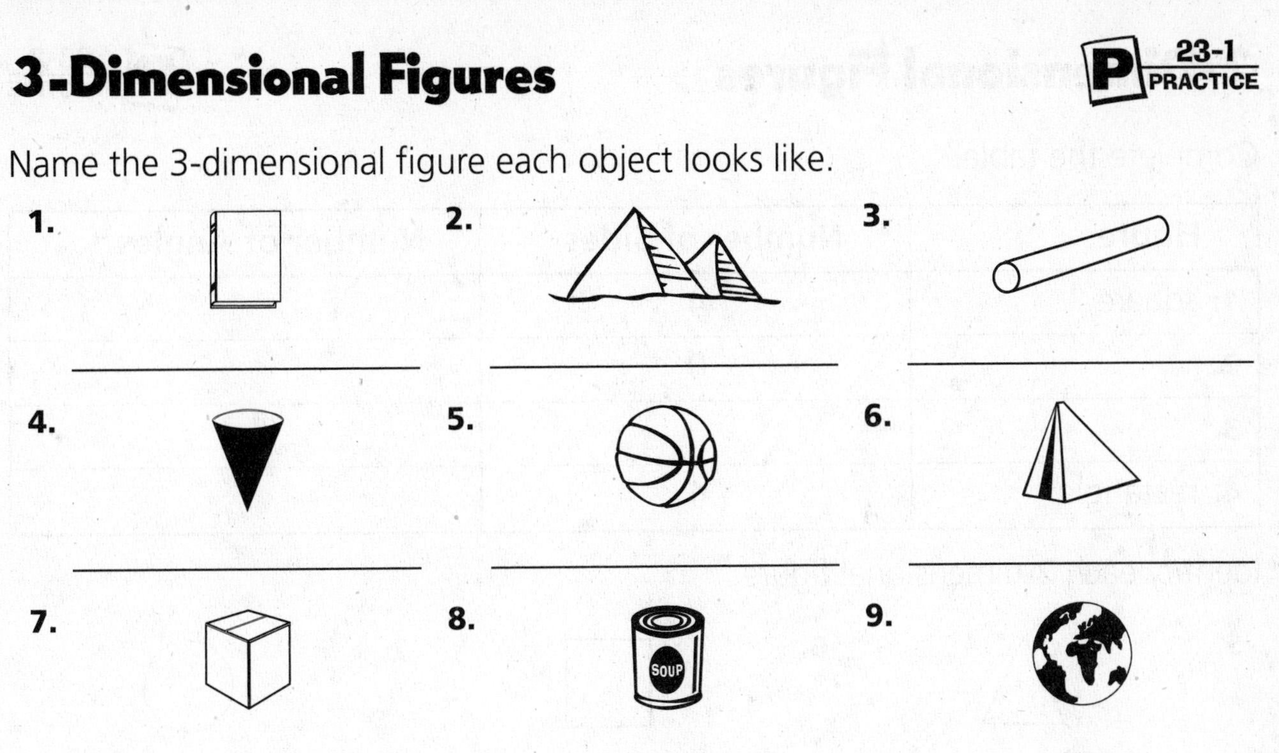

1. ____________ **2.** ____________ **3.** ____________

4. ____________ **5.** ____________ **6.** ____________

7. ____________ **8.** ____________ **9.** ____________

10. Identify the figures that were used to build this house.

____

____

11. Copy and fold. Identify the 3-dimensional figure.

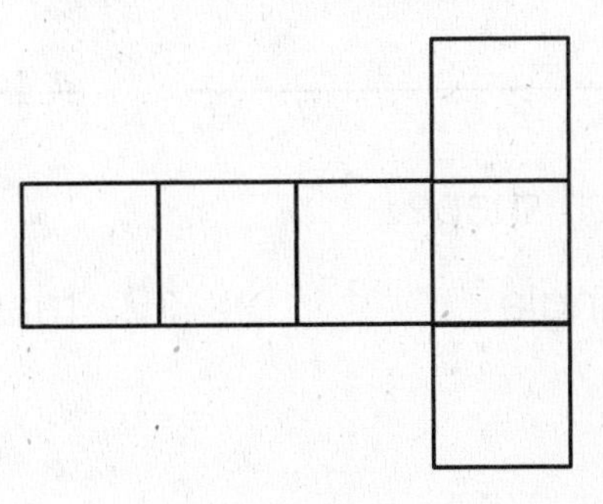

____

Complete the chart.

Figure	Number of Faces	Number of Edges	Number of Vertices
12.	6	____	____
13.	____	8	____

Name ________________________________

2-Dimensional Figures

23-2 PRACTICE

Complete the table.

Figure	Number of Sides	Number of Angles
1. square	4	
2.	0	
3.		3
4. rectangle	4	

Identify each 2-dimensional figure.

5.  ____________ **6.** ____________ **7.** ____________

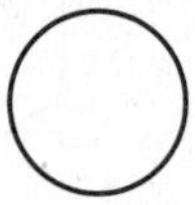

8. 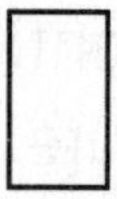 ____________ **9.** 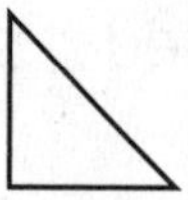 ____________ **10.** ____________

Draw each shape.

11. rectangle **12.** triangle **13.** square

Problem Solving

Solve.

14. Look at the shape of this worksheet page. What 2-dimensional figure is it like?

____

15. Suppose you want to trace a circle on a sheet of paper. Which of these could you use to help you: the bottom of a soup can or the bottom of a cereal box?

____

Use with Grade 3, Chapter 23, Lesson 2, pages 506–507.

Name ____________________

Lines, Line Segments, Rays, and Angles

Name each figure. Choose from the box below.

angle	right angle	circle	vertical line	radius
line	line segment	parallel lines	intersecting lines	

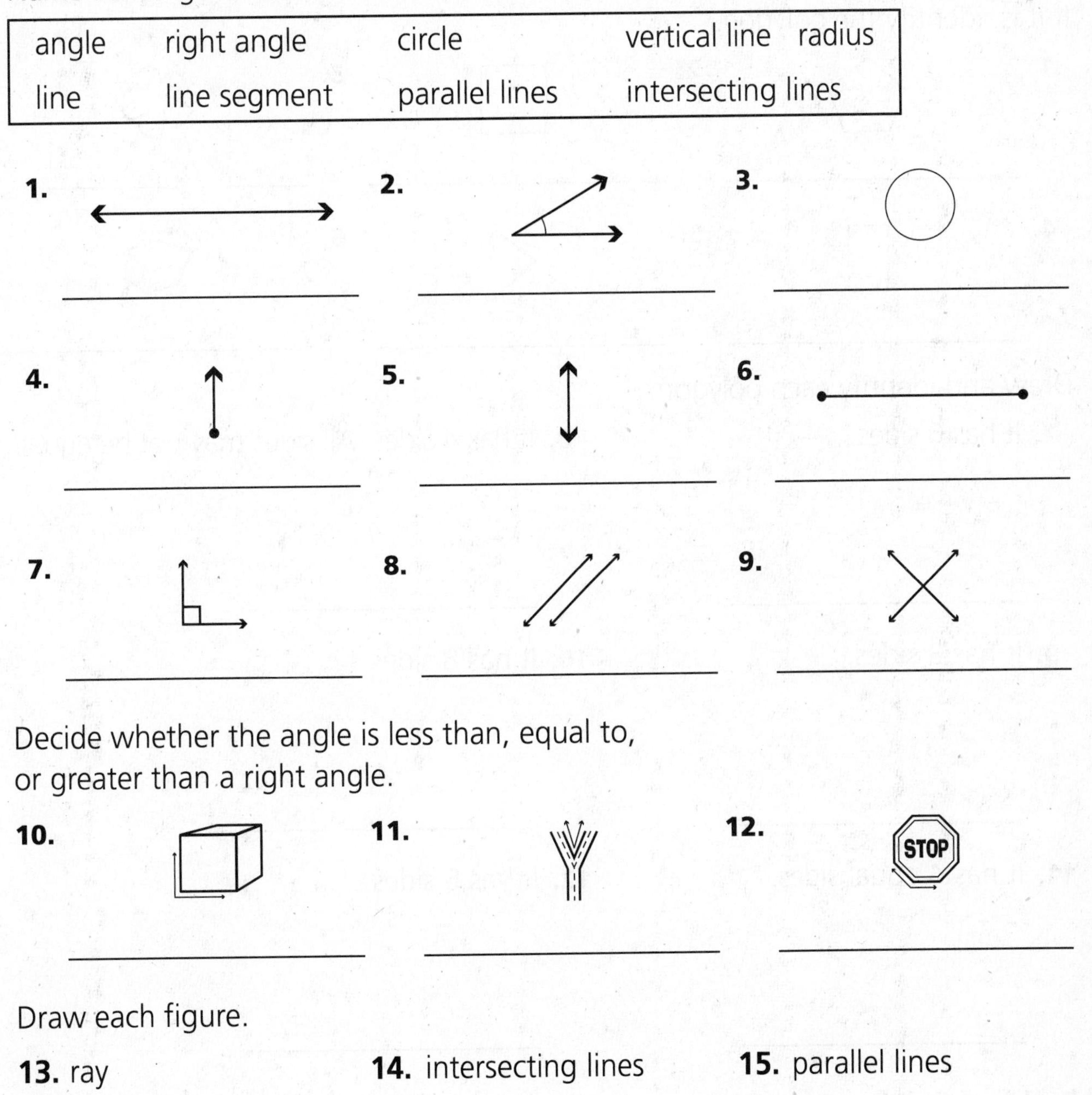

1. ____________________
2. ____________________
3. ____________________
4. ____________________
5. ____________________
6. ____________________
7. ____________________
8. ____________________
9. ____________________

Decide whether the angle is less than, equal to, or greater than a right angle.

10. ____________________
11. ____________________
12. ____________________

Draw each figure.

13. ray

14. intersecting lines

15. parallel lines

Name ______________________________

Polygons

P 23-4 PRACTICE

Write *yes* or *no* to tell if each figure is a polygon.
If it is, identify the polygon.

1. ______________

2. ______________

3. ______________

4. ______________

5. ______________

6. ______________

Draw and identify each polygon.

7. It has 6 sides.

____

8. It has 4 sides. All sides may not be equal.

____

9. It has 3 sides.

____

10. It has 8 sides.

____

11. It has 4 equal sides.

____

12. It has 5 sides.

____

Problem Solving

Solve.

13. The library at Ladew Mansion in Maryland has 8 sides. What is the shape of the library?

____

14. A kitchen tile has 4 equal sides. What is the shape of the tile?

____

Name ___________________________

Triangles

23-5 PRACTICE

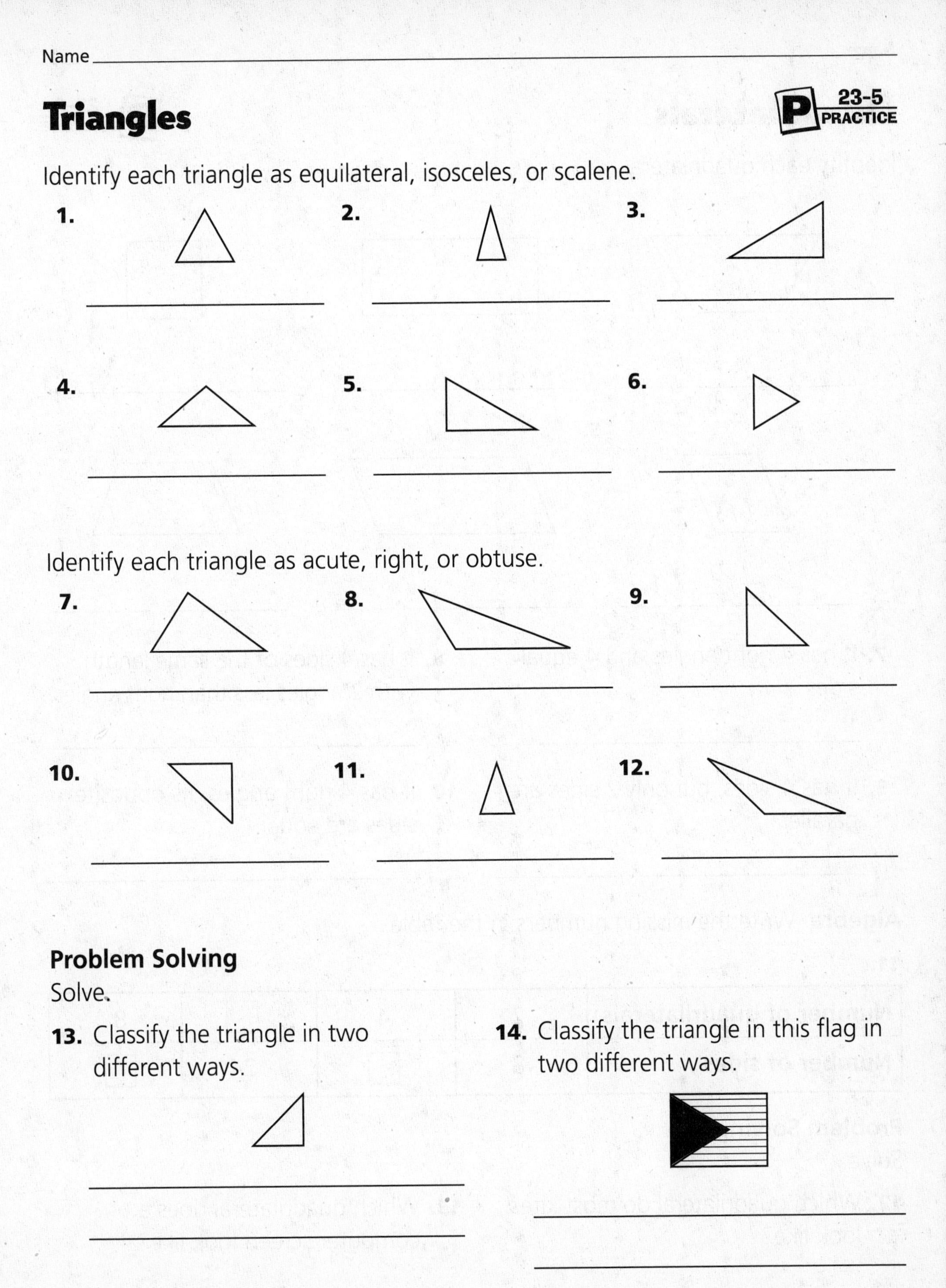

Identify each triangle as equilateral, isosceles, or scalene.

1. ______________

2. ______________

3. ______________

4. ______________

5. ______________

6. ______________

Identify each triangle as acute, right, or obtuse.

7. ______________

8. ______________

9. ______________

10. ______________

11. ______________

12. ______________

Problem Solving

Solve.

13. Classify the triangle in two different ways.

____

____

14. Classify the triangle in this flag in two different ways.

____

____

Name ______________________________

Quadrilaterals

P 23-6 PRACTICE

Identify each quadrilateral.

1. ______________

2. ______________

3. ______________

4. ______________

5. ______________

6. ______________

7. It has 4 right angles and 4 equal sides.

____

8. It has 4 sides of the same length with 2 angles less than right angles.

____

9. It has 4 sides, but only 2 sides are parallel.

____

10. It has 4 right angles. Its opposite sides are equal.

____

Algebra Write the missing numbers in the table.

11.

Number of quadrilaterals	2	4	☐	8
Number of sides	8	☐	24	☐

Problem Solving

Solve.

12. Which quadrilateral do most kites look like?

____

13. Which quadrilateral does a computer screen look like?

____

Name ______________________________

Problem Solving: Skill

Use a Diagram

Use data from the illustration to solve problems 1 and 2.

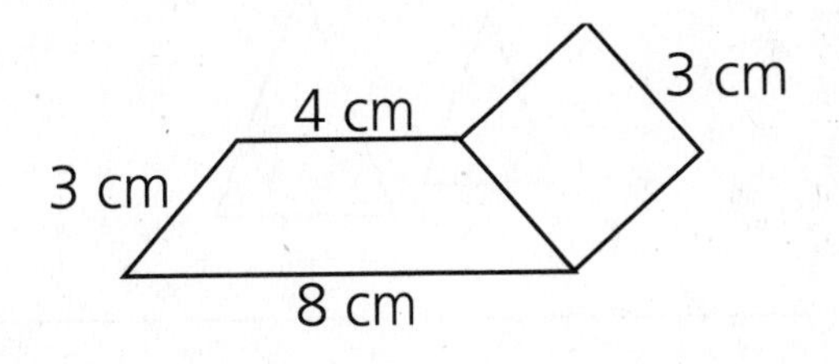

1. What are the lengths of the unlabeled sides of the figure?

____

2. How did you find the unlabeled sides of the figure?

____

____

____

Use data from the illustration to solve problems 3 and 4.

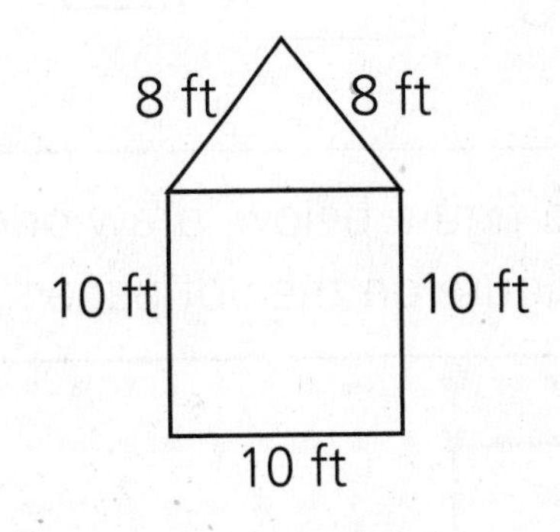

3. Linda designed this room. What two polygons make up this room?

____

4. What label is missing from the diagram?

____

Mixed Strategy Review

Solve.

5. Carly has a piece of ribbon that is 30 inches long. If she cuts off two pieces that are each 9 inches long, how much of the ribbon is left?

____

6. Matt drew a diagram of a triangular garden in which one angle was greater than a right angle. How would you classify this shape?

____

Name ____________________

Congruent and Similar Figures

P 24-1 PRACTICE

Write whether the figures are congruent, similar, or neither.

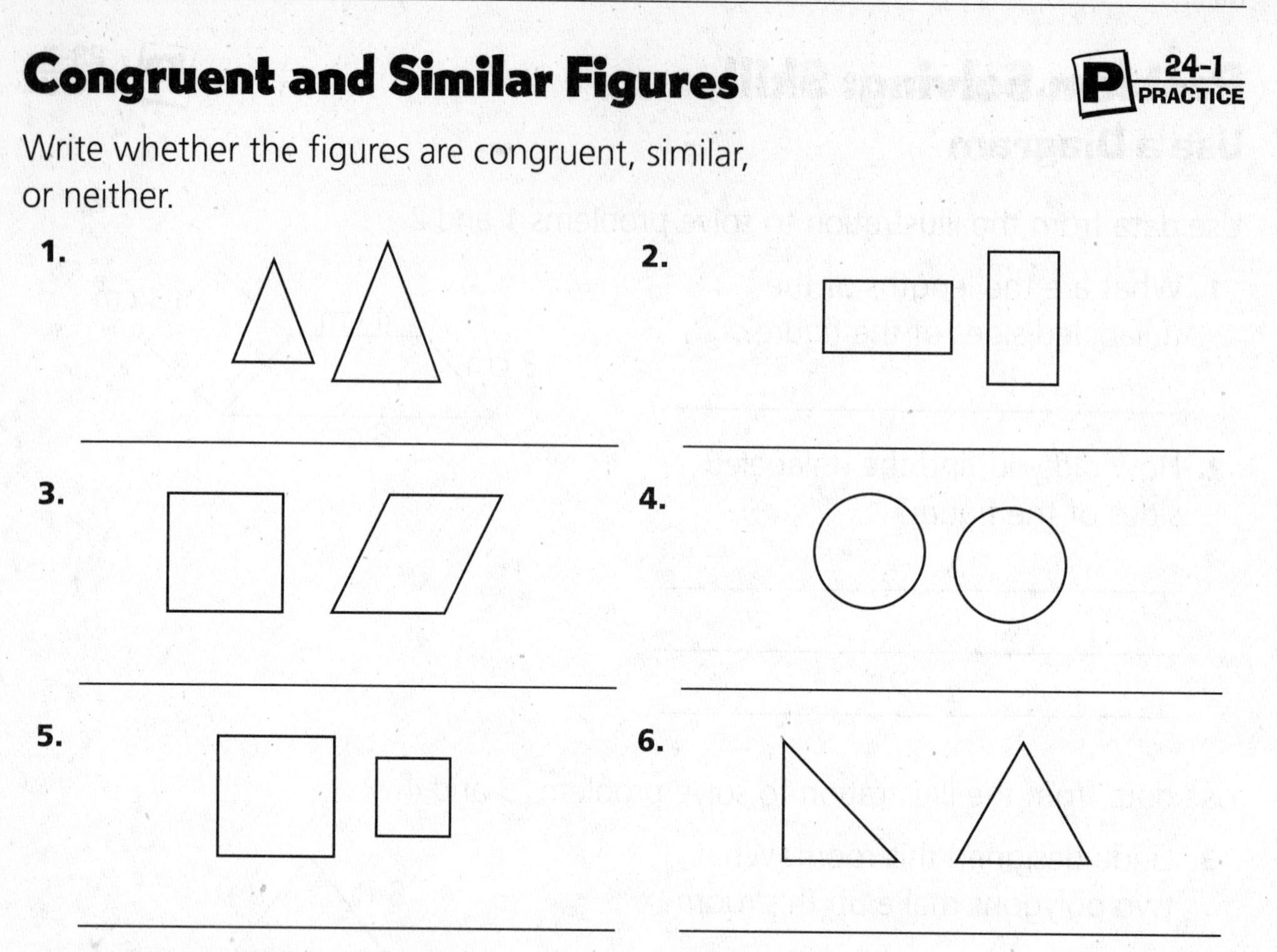

For each figure below, draw one congruent figure and one similar figure on the dot paper.

7.

8.

Problem Solving

Solve.

9. Are all the doors in your school similar? Are they congruent?

____

10. Picture the dinner plates in your home. Are all the plates similar? Are they congruent?

____

Name ____________________

Write reflection, rotation, or translation to describe how each figure was moved.

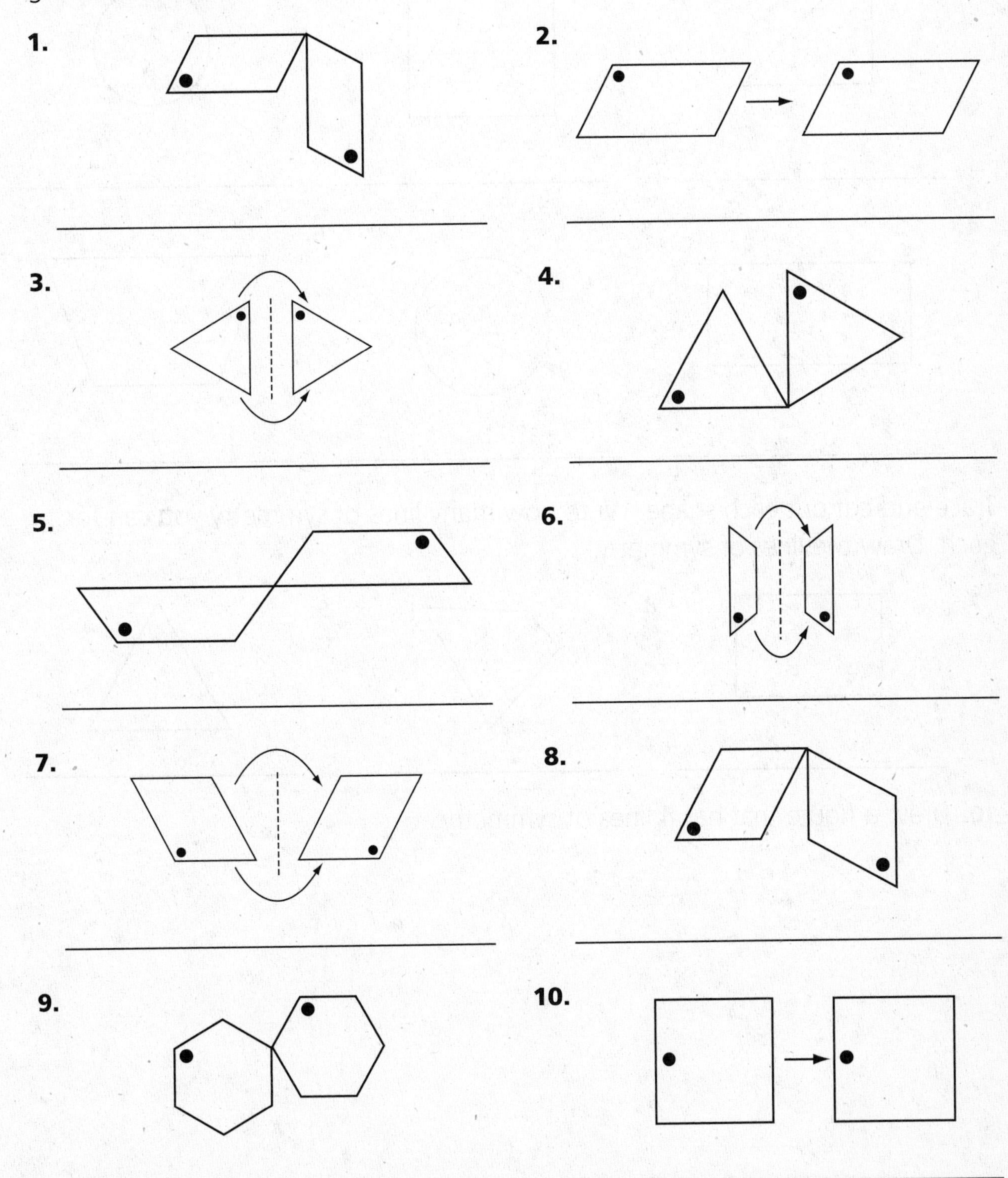

Name ______________________________

Explore Symmetry

P 24-3 PRACTICE

Write *yes* or *no* to tell if each line is a line of symmetry.

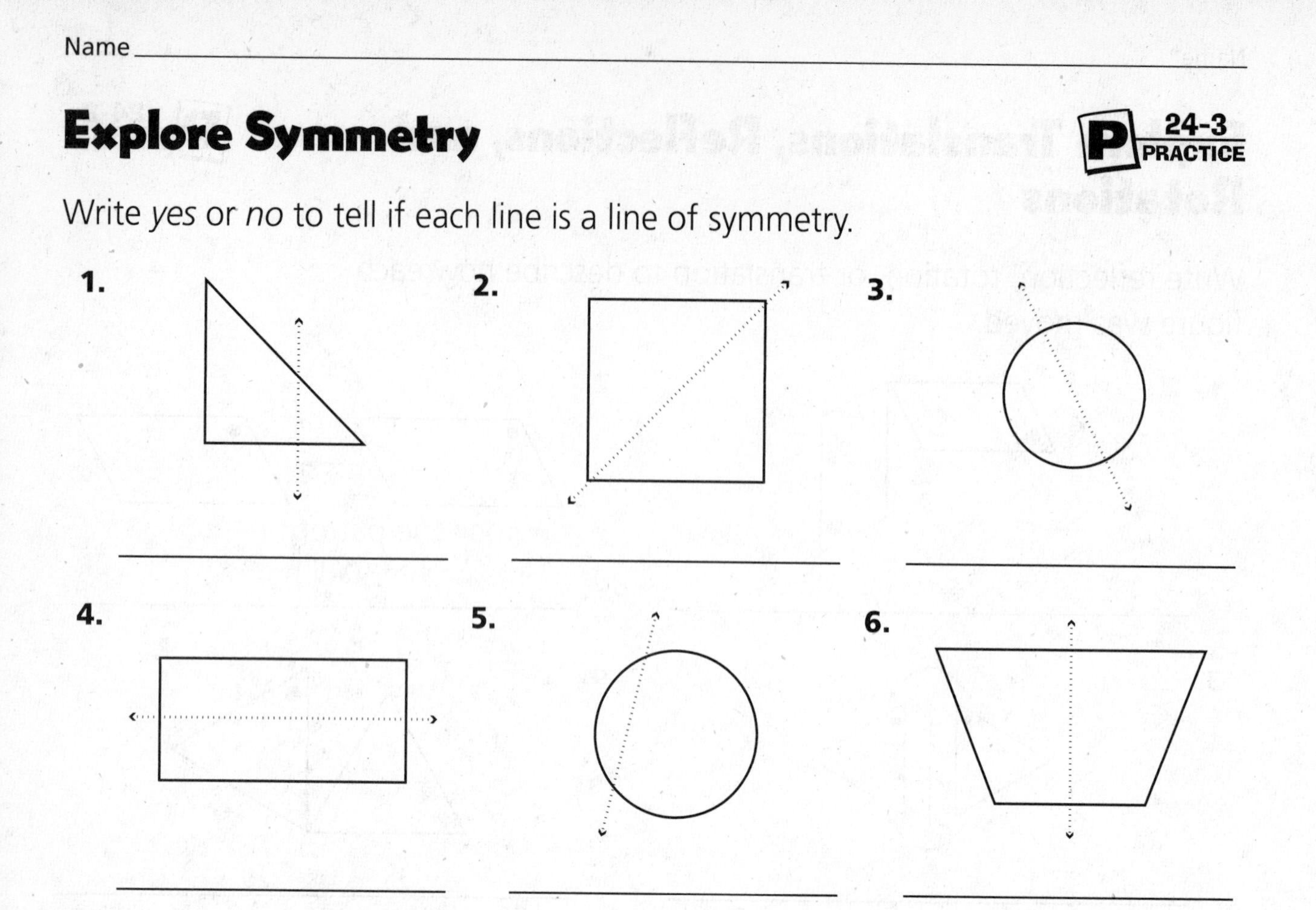

1. ____________ 2. ____________ 3. ____________

4. ____________ 5. ____________ 6. ____________

Trace and cut out each shape. Write how many lines of symmetry you can for each. Draw the lines of symmetry.

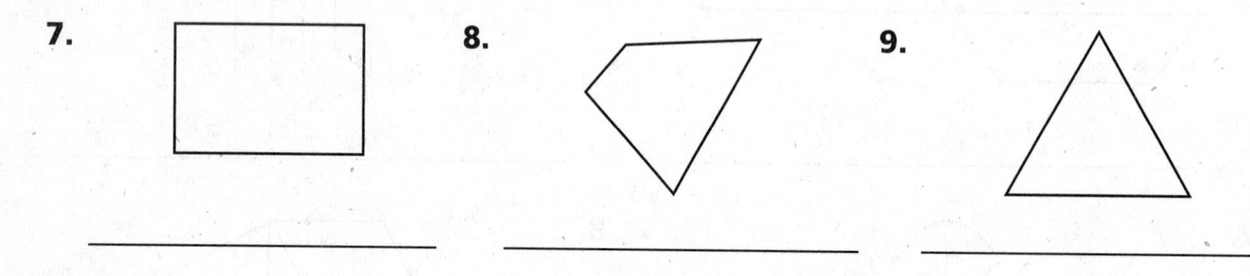

7. ____________ 8. ____________ 9. ____________

10. Draw a figure that has 4 lines of symmetry.

Name ____

Problem Solving: Strategy

Find a Pattern

P 24-4 PRACTICE

Find a pattern to solve.

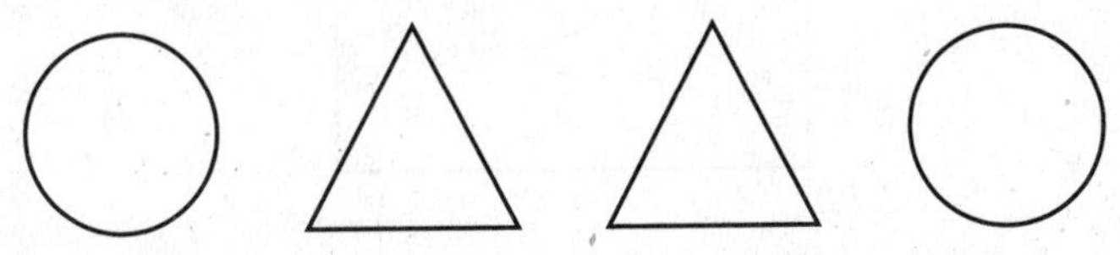
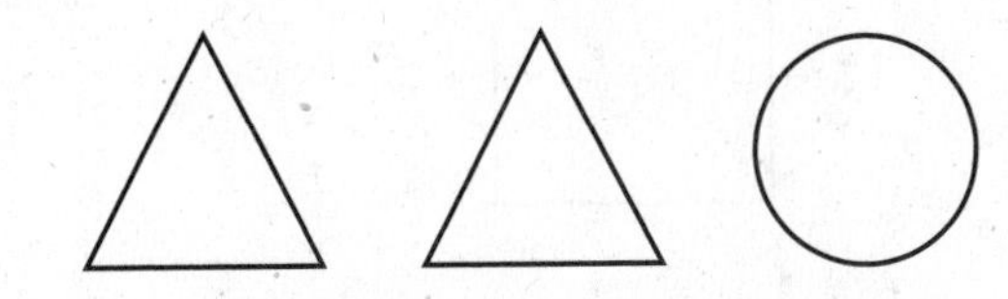

1. Denise paints the pattern shown above on a wall hanging. What shape do you think she paints next?

____

2. How can you use numbers to describe the pattern that Denise paints on her wall hanging?

____

3. Mark is making a pattern that repeats 4 birds and 2 cats. If he draws 36 animals, how many birds will he draw? How many cats?

____

4. Tawana is using tiles to make a design. The first row has 2 red tiles and 18 black tiles. The second has 4 red and 16 black. The third row has 6 red and 14 black. If this pattern continues, how many black tiles will be in the sixth row?

____

Mixed Strategy Review

5. Felix, Ron, Liz, and Tina buy White House souvenirs. One buys a model house, one buys a cap, one buys a shirt, and one buys a book. Felix and Ron buy souvenirs that can be worn. Tina does not buy a book. Who buys a model house?

____

6. **Write a problem** which you could solve by finding a pattern. Share it with others.

____

____

____

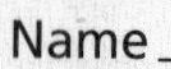

Name ____________________

Perimeter

P 24-5 PRACTICE

Use a centimeter ruler to find each perimeter.

1.

____

2.

____

3.

____

4.

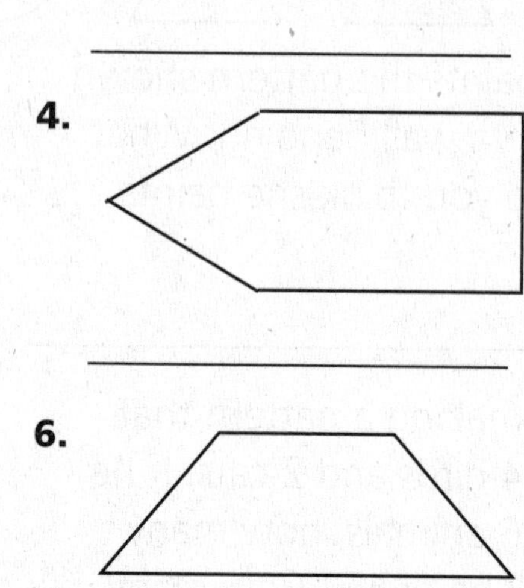

____

5.

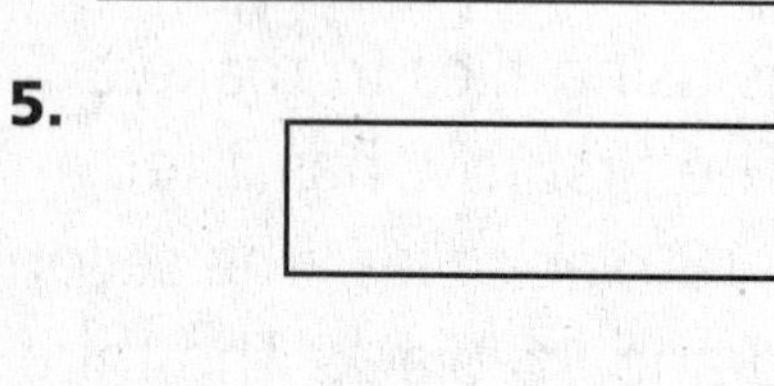

____

6.

____

Algebra Write the number that makes each sentence true.

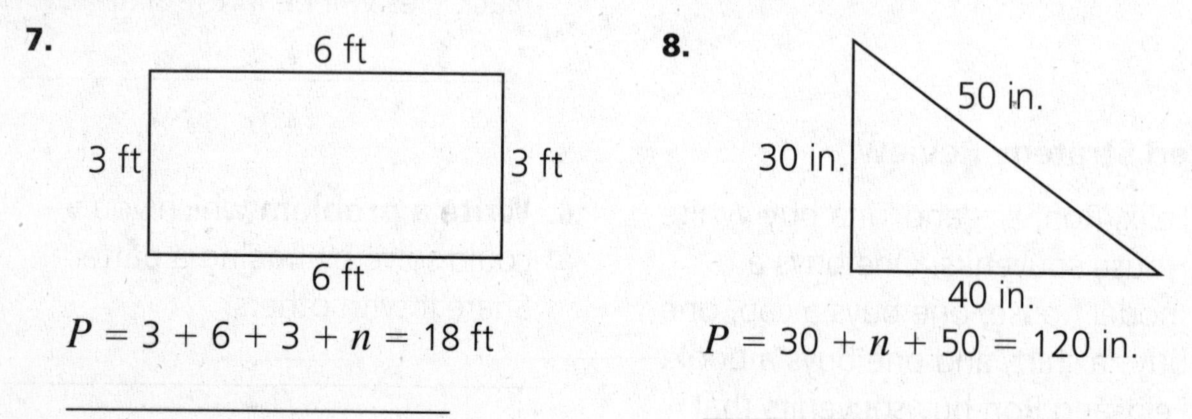

7. $P = 3 + 6 + 3 + n = 18$ ft

____

8. $P = 30 + n + 50 = 120$ in.

____

Problem Solving

Solve.

9. Kyle's garden is 15 ft long and 8 ft wide. He wants to put a fence around it. How much fencing will Kyle need?

____

10. Bing builds a fish pond in his back yard. The pond measures 121 in. wide and 140 in. long. What is the perimeter of the fish pond?

____

Name ______________________________

Area

Find each area in square units.

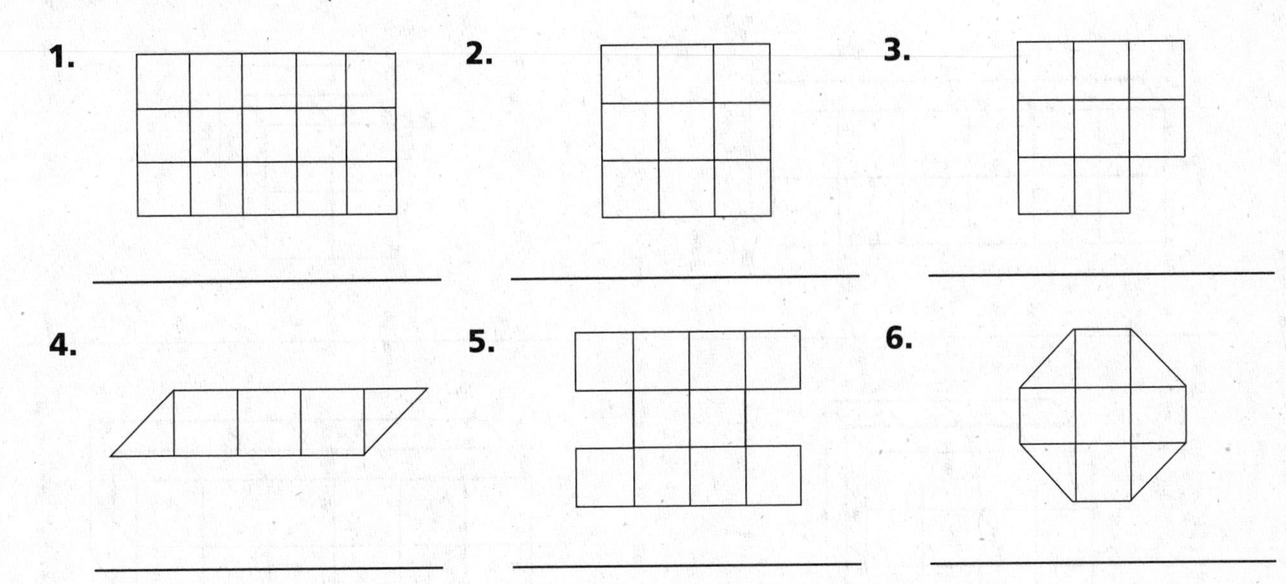

Draw a figure with the given area. Use the graph paper below.

7. 12 square units

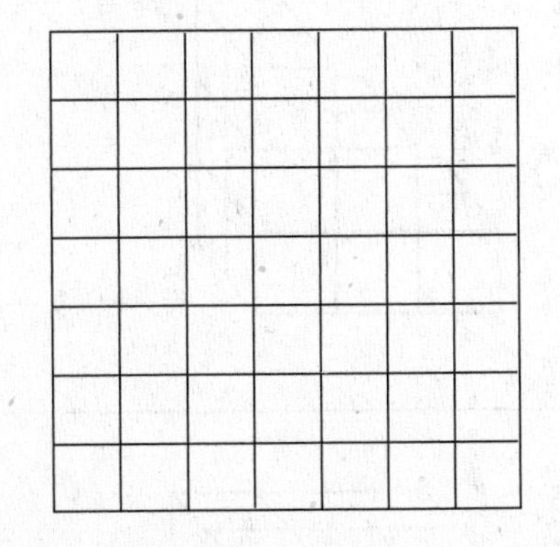

8. 18 square units

9. 25 square units

10. 30 square units

Name ______________________________

Explore Volume

Find each volume in cubic units.

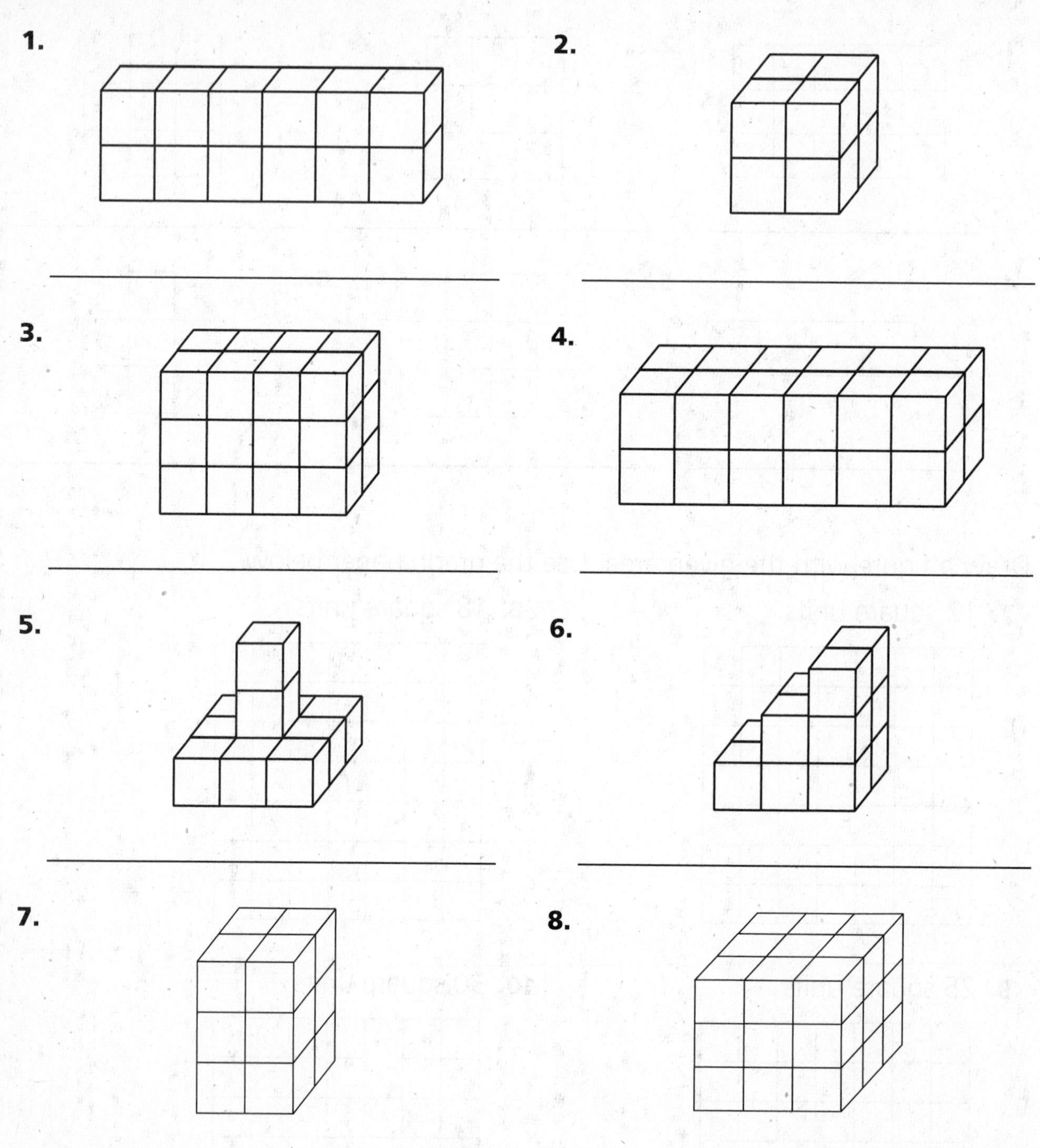

Name ______________________________

Parts of a Whole

P 25-1 PRACTICE

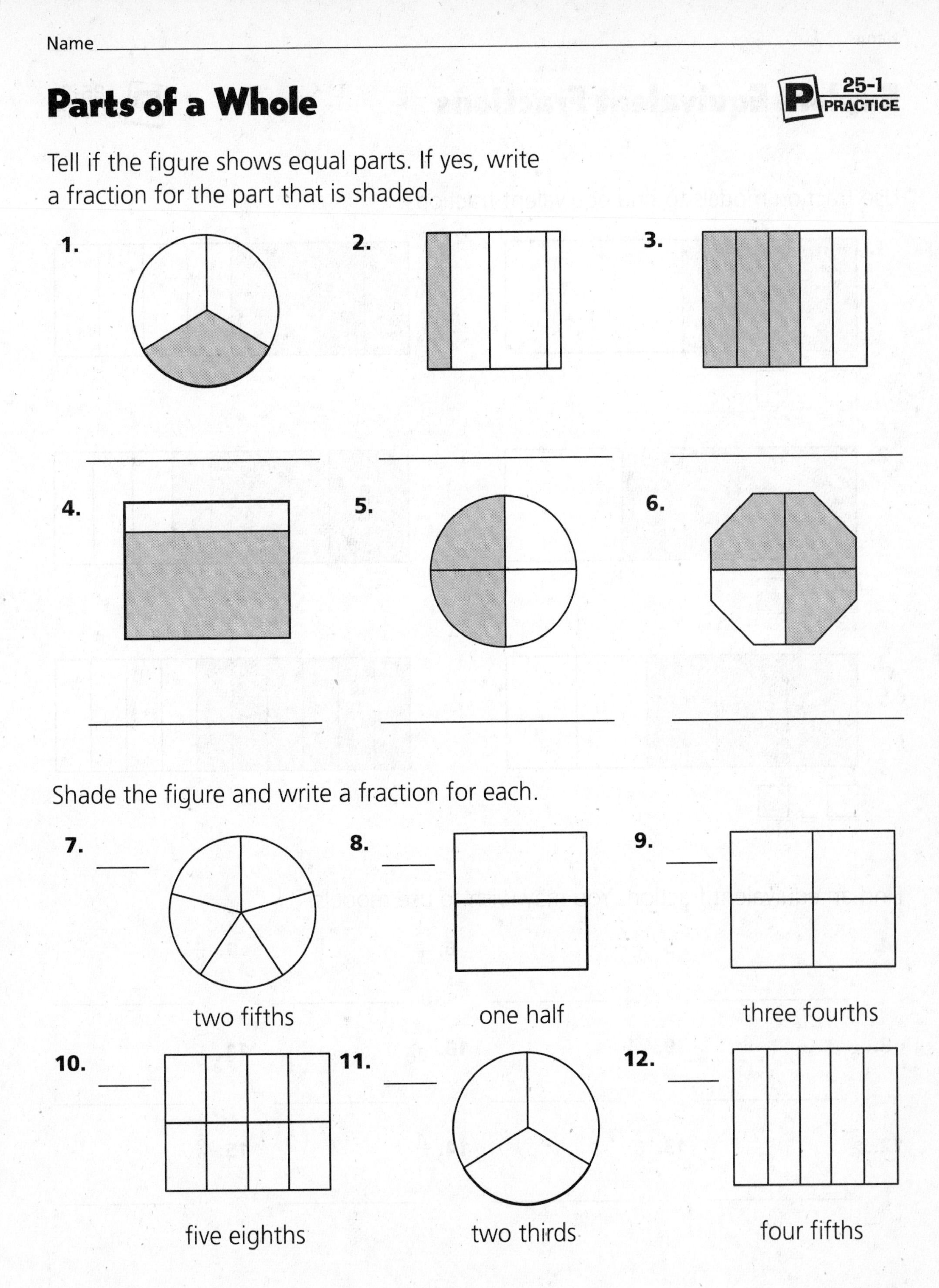

Tell if the figure shows equal parts. If yes, write a fraction for the part that is shaded.

1. ______________

2. ______________

3. ______________

4. ______________

5. ______________

6. ______________

Shade the figure and write a fraction for each.

7. ___ two fifths

8. ___ one half

9. ___ three fourths

10. ___ five eighths

11. ___ two thirds

12. ___ four fifths

Name ____________________

Explore Equivalent Fractions

Use fraction models to find equivalent fractions.

1.

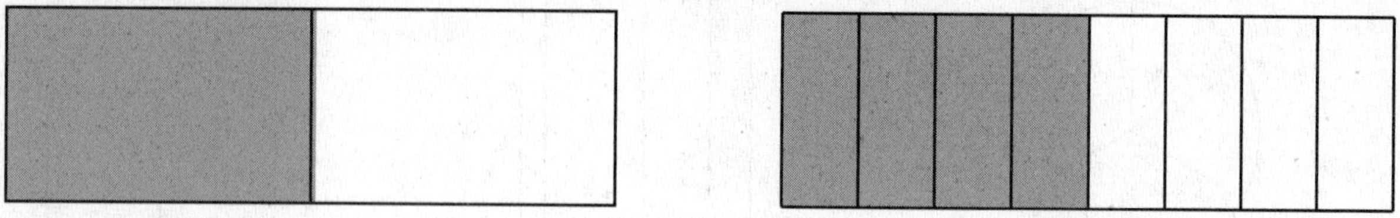

$\frac{\square}{2} = \frac{\square}{8}$

2.

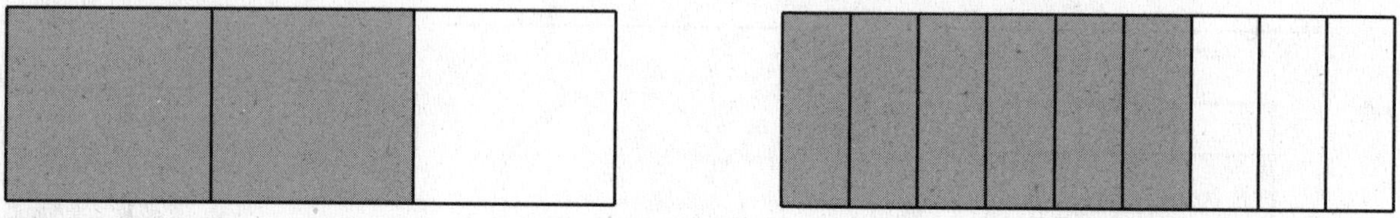

$\frac{\square}{3} = \frac{\square}{9}$

3.

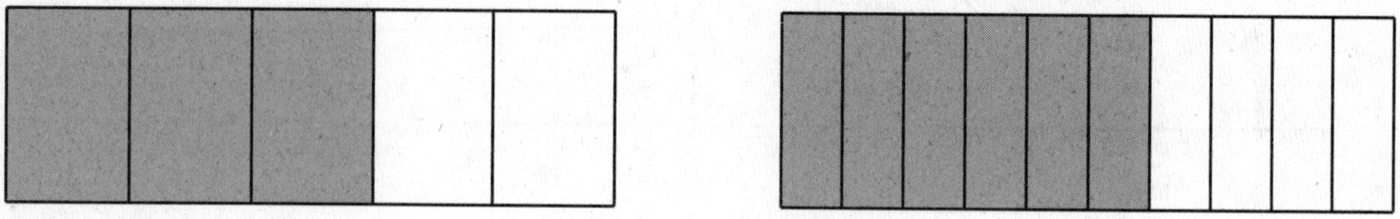

$\frac{\square}{5} = \frac{\square}{10}$

Find an equivalent fraction. You may wish to use models.

4. $\frac{1}{3}$ ________

5. $\frac{3}{8}$ ________

6. $\frac{2}{5}$ ________

7. $\frac{4}{8}$ ________

8. $\frac{1}{6}$ ________

9. $\frac{4}{12}$ ________

10. $\frac{2}{10}$ ________

11. $\frac{3}{5}$ ________

12. $\frac{3}{6}$ ________

13. $\frac{2}{8}$ ________

14. $\frac{4}{5}$ ________

15. $\frac{6}{8}$ ________

Use with Grade 3, Chapter 25, Lesson 2, pages 558–559.

Name ______________________________

Fractions in Simplest Form

P 25-3 PRACTICE

Write each fraction in simplest form.

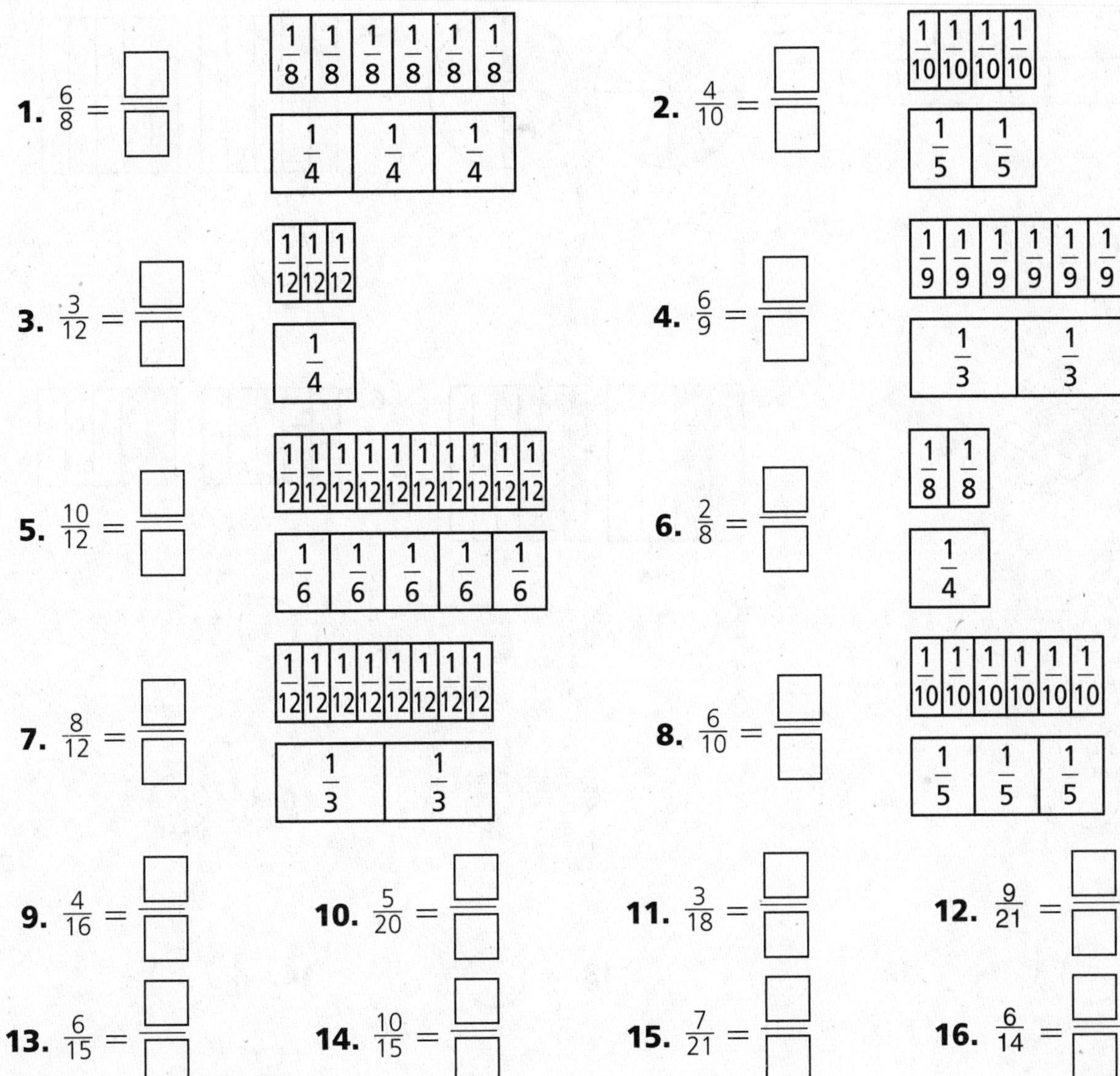

9. $\frac{4}{16} = \frac{\square}{\square}$ **10.** $\frac{5}{20} = \frac{\square}{\square}$ **11.** $\frac{3}{18} = \frac{\square}{\square}$ **12.** $\frac{9}{21} = \frac{\square}{\square}$

13. $\frac{6}{15} = \frac{\square}{\square}$ **14.** $\frac{10}{15} = \frac{\square}{\square}$ **15.** $\frac{7}{21} = \frac{\square}{\square}$ **16.** $\frac{6}{14} = \frac{\square}{\square}$

Problem Solving

Solve.

17. Kristen cuts a cake into 10 pieces. She serves 4 pieces of the cake. How much of the cake does she serve? What is this fraction in simplest form?

____

18. Luke divides his garden into 16 equal square sections. He plants flowers in 8 of the squares. How can he write the squares with flowers as a fraction in simplest form?

____

Name ______________________________

Compare and Order Fractions

25-4 PRACTICE

Compare. Write >, <, or =.

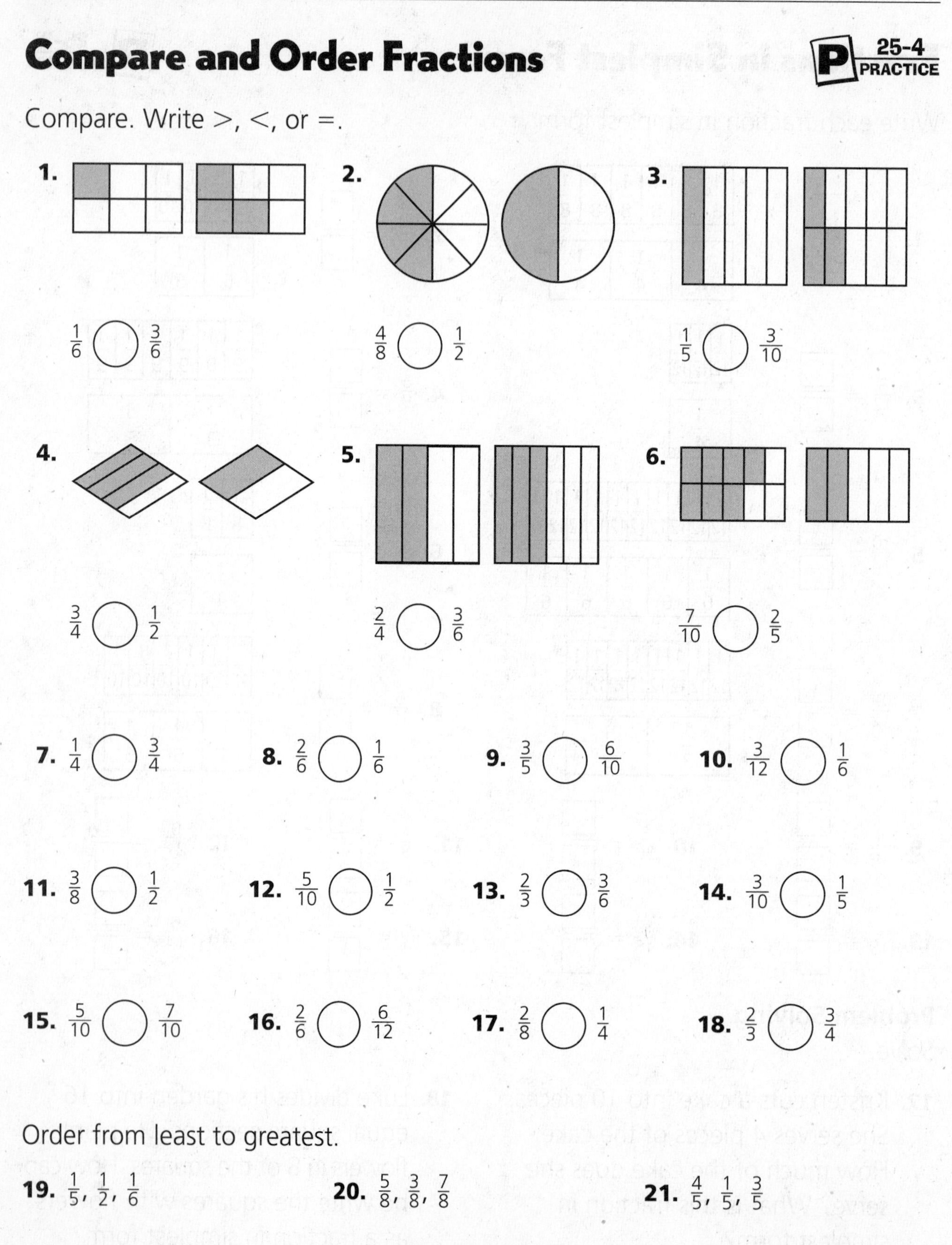

1. $\frac{1}{6}$ ◯ $\frac{3}{6}$

2. $\frac{4}{8}$ ◯ $\frac{1}{2}$

3. $\frac{1}{5}$ ◯ $\frac{3}{10}$

4. $\frac{3}{4}$ ◯ $\frac{1}{2}$

5. $\frac{2}{4}$ ◯ $\frac{3}{6}$

6. $\frac{7}{10}$ ◯ $\frac{2}{5}$

7. $\frac{1}{4}$ ◯ $\frac{3}{4}$

8. $\frac{2}{6}$ ◯ $\frac{1}{6}$

9. $\frac{3}{5}$ ◯ $\frac{6}{10}$

10. $\frac{3}{12}$ ◯ $\frac{1}{6}$

11. $\frac{3}{8}$ ◯ $\frac{1}{2}$

12. $\frac{5}{10}$ ◯ $\frac{1}{2}$

13. $\frac{2}{3}$ ◯ $\frac{3}{6}$

14. $\frac{3}{10}$ ◯ $\frac{1}{5}$

15. $\frac{5}{10}$ ◯ $\frac{7}{10}$

16. $\frac{2}{6}$ ◯ $\frac{6}{12}$

17. $\frac{2}{8}$ ◯ $\frac{1}{4}$

18. $\frac{2}{3}$ ◯ $\frac{3}{4}$

Order from least to greatest.

19. $\frac{1}{5}, \frac{1}{2}, \frac{1}{6}$ ______________________

20. $\frac{5}{8}, \frac{3}{8}, \frac{7}{8}$ ______________________

21. $\frac{4}{5}, \frac{1}{5}, \frac{3}{5}$ ______________________

Name ____

Parts of a Group

25-5 PRACTICE

Write a fraction for the part of each group that is shaded.

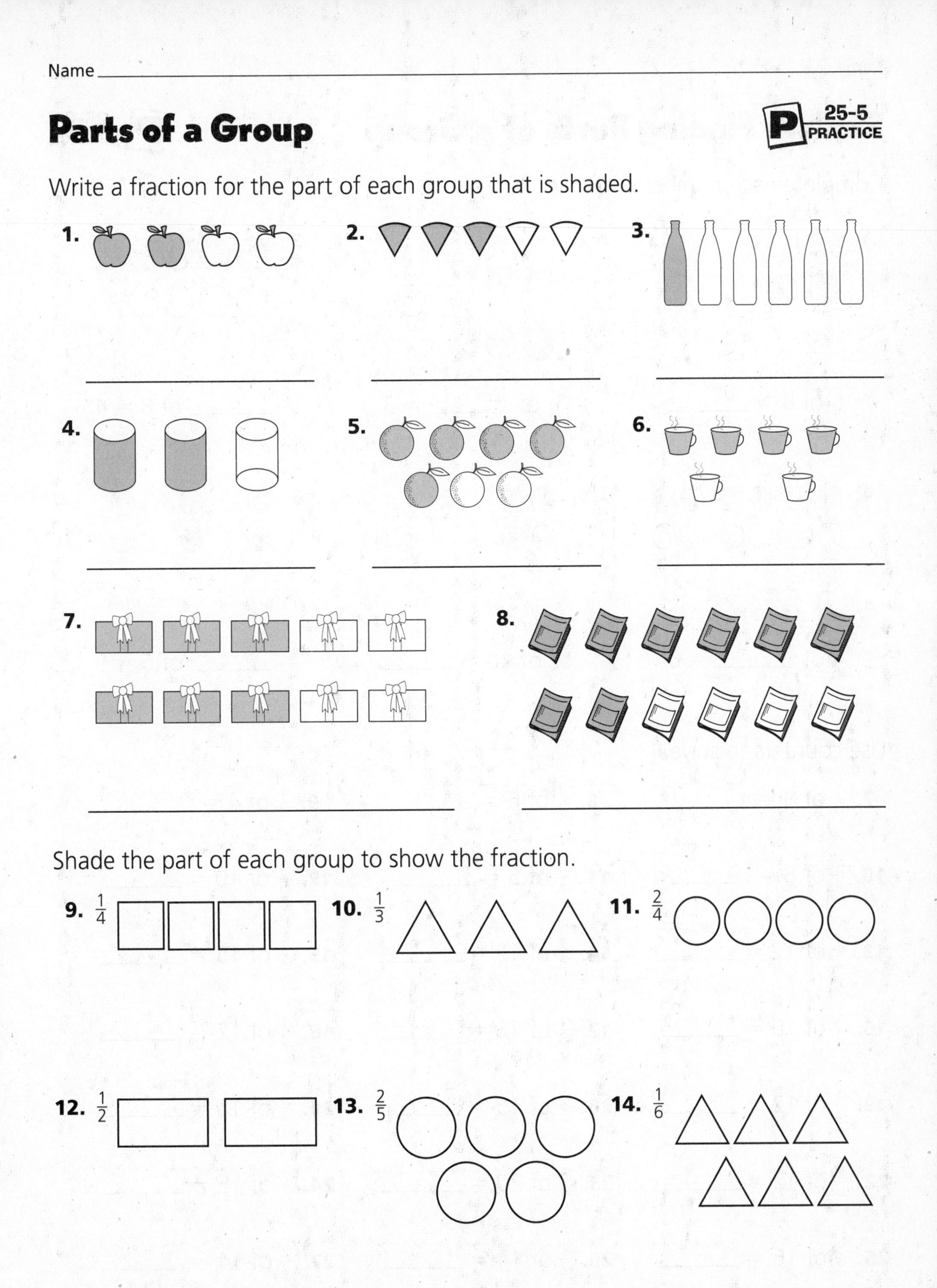

Shade the part of each group to show the fraction.

Name ______________________________

Explore Finding Parts of a Group

Complete each number sentence.

1. 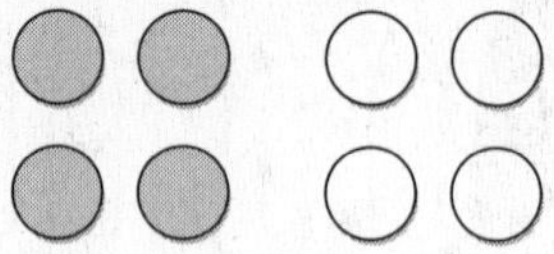 $\frac{2}{4}$ of 8 = ______

2. 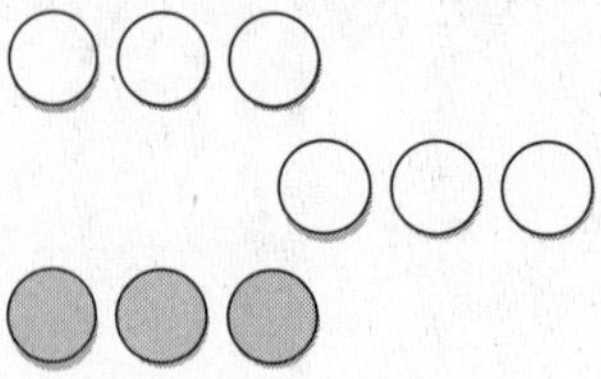 $\frac{1}{3}$ of ______ = 3

3. 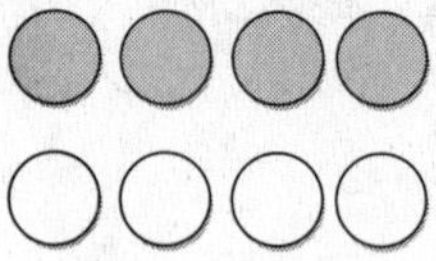 ______ of 8 = 4

4. 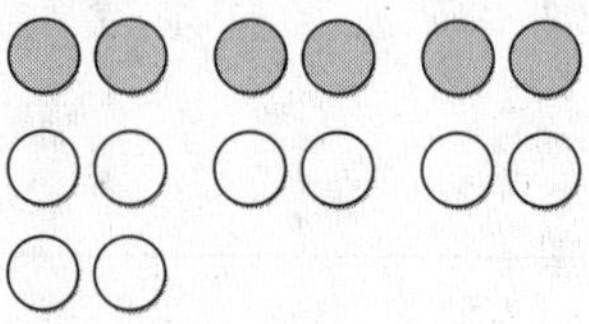 $\frac{3}{7}$ of ______ = 6

5. 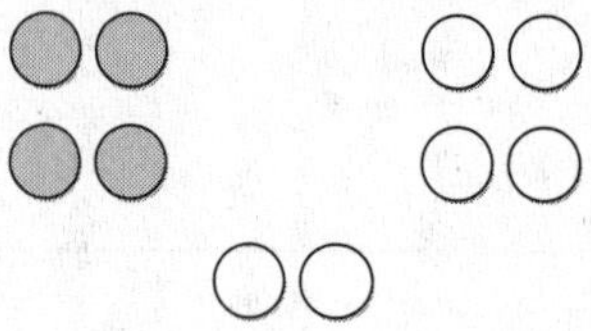 $\frac{2}{5}$ of 10 = ______

6. 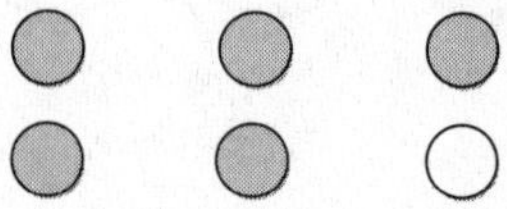 ______ of 6 = 5

Use counters to solve.

7. $\frac{1}{8}$ of 8 = ______
8. $\frac{2}{3}$ of 6 = ______
9. $\frac{1}{2}$ of 4 = ______
10. $\frac{3}{4}$ of 8 = ______
11. $\frac{4}{5}$ of 5 = ______
12. $\frac{3}{5}$ of 10 = ______
13. $\frac{1}{2}$ of 12 = ______
14. $\frac{2}{3}$ of 12 = ______
15. $\frac{1}{3}$ of 15 = ______
16. $\frac{1}{2}$ of 10 = ______
17. $\frac{1}{8}$ of 16 = ______
18. $\frac{1}{4}$ of 12 = ______
19. $\frac{1}{3}$ of 12 = ______
20. $\frac{3}{5}$ of 15 = ______
21. $\frac{3}{4}$ of 16 = ______
22. $\frac{5}{6}$ of 12 = ______
23. $\frac{3}{4}$ of 20 = ______
24. $\frac{2}{3}$ of 15 = ______
25. $\frac{5}{8}$ of 16 = ______
26. $\frac{1}{2}$ of 14 = ______
27. $\frac{2}{7}$ of 14 = ______

Use with Grade 3, Chapter 25, Lesson 6, pages 568–569.

Name ______________________________

Problem Solving: Skill
Check for Reasonableness

Solve. Is your answer reasonable?

1. The Middletown Farmer's Market has 24 booths. This Sunday, $\frac{1}{6}$ of the booths sell dairy products. Is it reasonable to say that 6 of the booths sell dairy products?

____

2. At the Middletown Farmer's Market, $\frac{1}{3}$ of the booths sell fruit. There are about 24 booths. Is it reasonable to say that only 4 booths sell fruit?

____

3. Nina buys 12 small boxes of raisins. She gives $\frac{2}{3}$ of the boxes to friends. Is it reasonable to say that she gives away 9 boxes?

____

4. Jerry makes 16 brownies. A customer buys $\frac{3}{4}$ of them. Is it reasonable to say that the customer buys 12 brownies?

____

Mixed Strategy Review

5. Ben buys a pie. He cuts the pie into 10 slices. Ben's friends eat $\frac{1}{2}$ of the pie. How many slices do they eat?

____

6. Lisa has 12 feet of paper. She uses $\frac{1}{3}$ of the paper to make a sign. How many feet of paper does Lisa use?

____

7. Denny makes 24 muffins. Of these muffins, $\frac{2}{3}$ are bran muffins. How many bran muffins does Denny make?

____

8. There are 16 students helping at a bake sale. Of these students, $\frac{1}{4}$ are in the third grade. How many third-grade students help at the bake sale?

____

Name ____________________

Mixed Numbers

26-1 PRACTICE

Write as a mixed number.

1. ____________

2. ____________

3. 1C 2/3 1/3 1C 2/3 1/3 ____________

4. ____________

5. crayon $\frac{1}{2}$ 1 $\frac{1}{2}$ 2 $\frac{1}{2}$ 3 $\frac{1}{2}$ ____________

Measure to the nearest $\frac{1}{2}$ inch.

6. ____________

7. ____________

Draw a model of each mixed number.
Then write each mixed number in words.

8. $1\frac{3}{4}$ ____________

9. $2\frac{1}{8}$ ____________

10. $3\frac{1}{2}$ ____________

11. $2\frac{2}{3}$ ____________

Name ____________________

Explore Adding Fractions

P 26-2 PRACTICE

Use fraction models to find the sum.

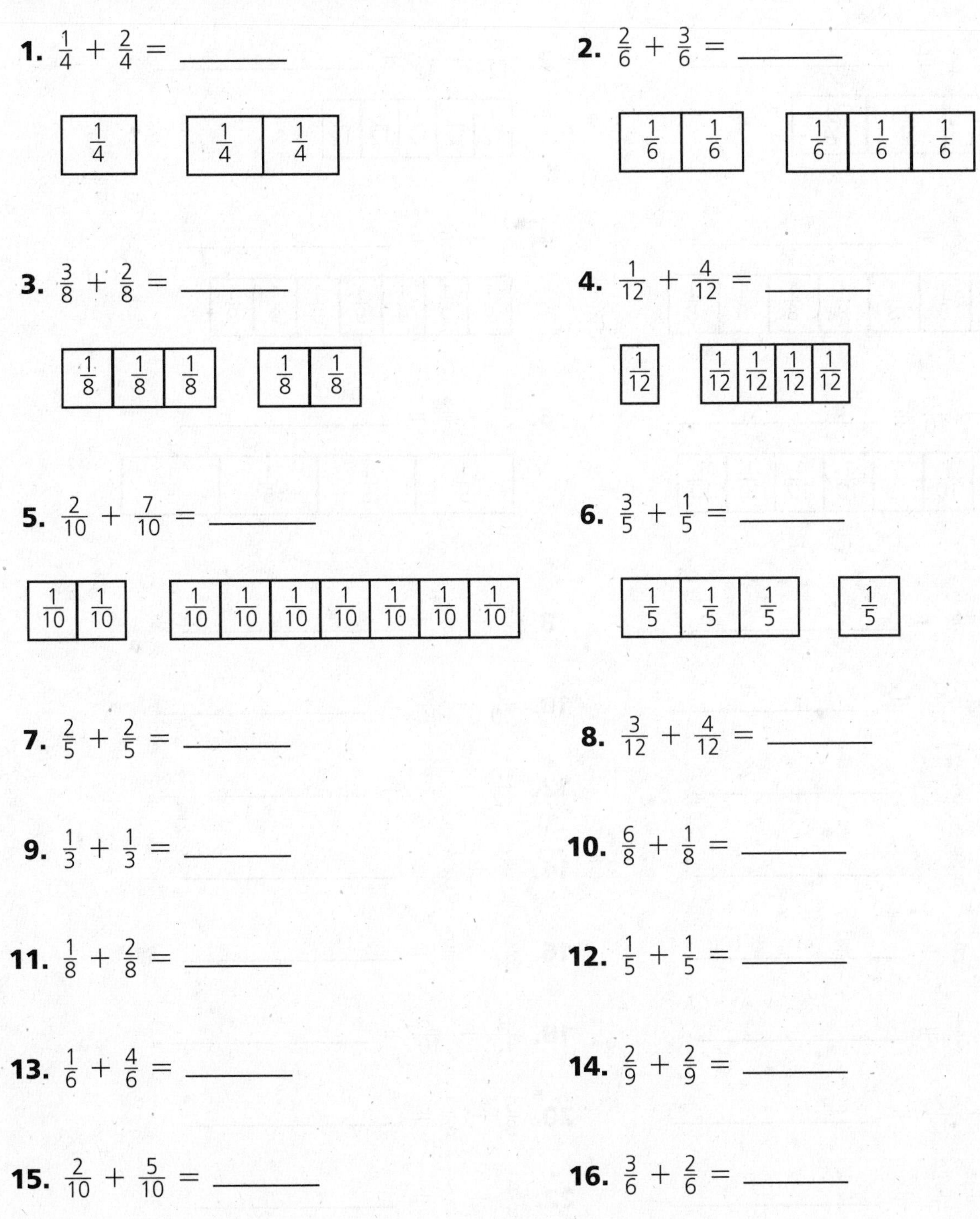

1. $\frac{1}{4} + \frac{2}{4} =$ ______

2. $\frac{2}{6} + \frac{3}{6} =$ ______

3. $\frac{3}{8} + \frac{2}{8} =$ ______

4. $\frac{1}{12} + \frac{4}{12} =$ ______

5. $\frac{2}{10} + \frac{7}{10} =$ ______

6. $\frac{3}{5} + \frac{1}{5} =$ ______

7. $\frac{2}{5} + \frac{2}{5} =$ ______

8. $\frac{3}{12} + \frac{4}{12} =$ ______

9. $\frac{1}{3} + \frac{1}{3} =$ ______

10. $\frac{6}{8} + \frac{1}{8} =$ ______

11. $\frac{1}{8} + \frac{2}{8} =$ ______

12. $\frac{1}{5} + \frac{1}{5} =$ ______

13. $\frac{1}{6} + \frac{4}{6} =$ ______

14. $\frac{2}{9} + \frac{2}{9} =$ ______

15. $\frac{2}{10} + \frac{5}{10} =$ ______

16. $\frac{3}{6} + \frac{2}{6} =$ ______

Name ______________________________

Explore Subtracting Fractions

Use fraction models to subtract.

1. $\frac{3}{4} - \frac{2}{4} =$ ______________

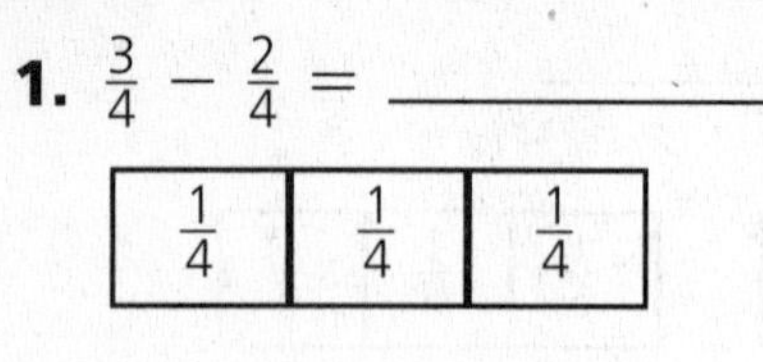

2. $\frac{5}{12} - \frac{4}{12} =$ ______________

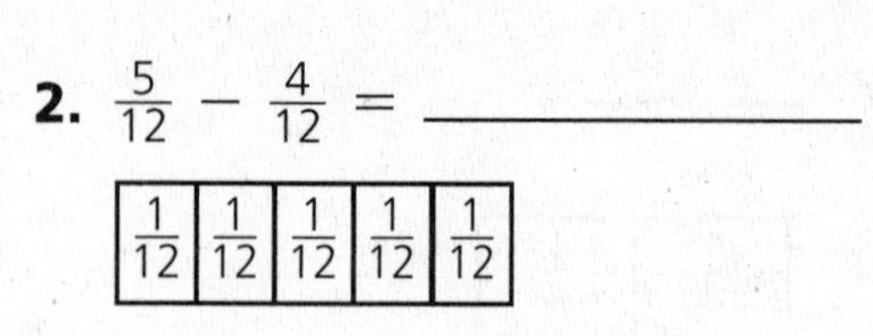

3. $\frac{7}{8} - \frac{2}{8} =$ ______________

$\frac{1}{8}$	$\frac{1}{8}$	$\frac{1}{8}$	$\frac{1}{8}$	$\frac{1}{8}$	$\frac{1}{8}$	$\frac{1}{8}$

4. $\frac{7}{9} - \frac{3}{9} =$ ______________

$\frac{1}{9}$	$\frac{1}{9}$	$\frac{1}{9}$	$\frac{1}{9}$	$\frac{1}{9}$	$\frac{1}{9}$	$\frac{1}{9}$

5. $\frac{7}{10} - \frac{4}{10} =$ ______________

$\frac{1}{10}$	$\frac{1}{10}$	$\frac{1}{10}$	$\frac{1}{10}$	$\frac{1}{10}$	$\frac{1}{10}$	$\frac{1}{10}$

6. $\frac{4}{5} - \frac{2}{5} =$ ______________

$\frac{1}{5}$	$\frac{1}{5}$	$\frac{1}{5}$	$\frac{1}{5}$

7. $\frac{5}{7} - \frac{2}{7} =$ ______________

8. $\frac{6}{8} - \frac{1}{8} =$ ______________

9. $\frac{7}{12} - \frac{2}{12} =$ ______________

10. $\frac{9}{10} - \frac{8}{10} =$ ______________

11. $\frac{4}{5} - \frac{1}{5} =$ ______________

12. $\frac{10}{12} - \frac{3}{12} =$ ______________

13. $\frac{6}{8} - \frac{3}{8} =$ ______________

14. $\frac{2}{3} - \frac{1}{3} =$ ______________

15. $\frac{8}{9} - \frac{3}{9} =$ ______________

16. $\frac{6}{8} - \frac{5}{8} =$ ______________

17. $\frac{3}{7} - \frac{1}{7} =$ ______________

18. $\frac{9}{10} - \frac{2}{10} =$ ______________

19. $\frac{3}{12} - \frac{2}{12} =$ ______________

20. $\frac{5}{8} - \frac{2}{8} =$ ______________

21. $\frac{8}{10} - \frac{5}{10} =$ ______________

22. $\frac{4}{9} - \frac{2}{9} =$ ______________

23. $\frac{8}{12} - \frac{3}{12} =$ ______________

24. $\frac{5}{6} - \frac{4}{6} =$ ______________

Name ____________________

Add and Subtract Fractions

Add or subtract. Write the answer in simplest form.

1. $\frac{1}{4} + \frac{3}{4} =$ ____________
2. $\frac{2}{8} + \frac{4}{8} =$ ____________
3. $\frac{4}{10} + \frac{3}{10} =$ ____________
4. $\frac{2}{3} + \frac{2}{3} =$ ____________
5. $\frac{3}{5} + \frac{1}{5} =$ ____________
6. $\frac{2}{5} + \frac{2}{5} =$ ____________
7. $\frac{6}{7} - \frac{3}{7} =$ ____________
8. $\frac{6}{8} - \frac{1}{8} =$ ____________
9. $\frac{10}{12} - \frac{3}{12} =$ ____________
10. $\frac{2}{4} - \frac{1}{4} =$ ____________
11. $\frac{11}{12} - \frac{9}{12} =$ ____________
12. $\frac{5}{11} - \frac{3}{11} =$ ____________
13. $\frac{5}{9} + \frac{2}{9} =$ ____________
14. $\frac{1}{4} + \frac{2}{4} =$ ____________
15. $\frac{3}{9} + \frac{3}{9} =$ ____________
16. $\frac{4}{12} + \frac{4}{12} =$ ____________
17. $\frac{4}{12} + \frac{5}{12} =$ ____________
18. $\frac{3}{8} + \frac{3}{8} =$ ____________
19. $\frac{7}{8} - \frac{5}{8} =$ ____________
20. $\frac{5}{6} - \frac{3}{6} =$ ____________
21. $\frac{12}{15} - \frac{5}{15} =$ ____________
22. $\frac{12}{18} - \frac{11}{18} =$ ____________
23. $\frac{9}{10} - \frac{7}{10} =$ ____________
24. $\frac{5}{12} - \frac{1}{12} =$ ____________

Algebra Compare. Write >, <, or =.

25. $\frac{5}{8} + \frac{3}{8}$ ◯ $\frac{6}{8} + \frac{1}{8}$
26. $\frac{7}{8} - \frac{3}{8}$ ◯ $\frac{3}{4} - \frac{1}{4}$
27. $\frac{5}{9} - \frac{3}{9}$ ◯ $\frac{2}{3} - \frac{1}{3}$

Name ____________________

Probability

Write *certain*, *likely*, *unlikely*, or *impossible* to describe the probability.

1. Land on a 4. ____________

2. Land on a 1. ____________

3. Land on a 5. ____________

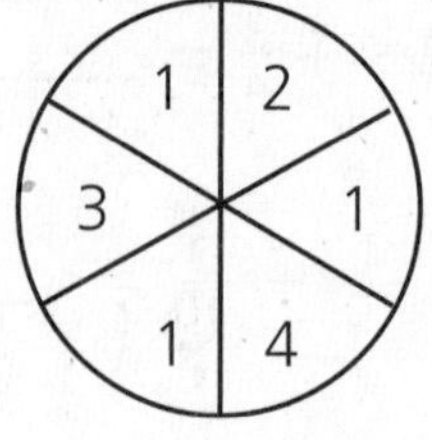

4. Pick an apple. ____________

5. Pick a banana. ____________

6. Pick a cherry. ____________

7. Pick an odd number. ____________

8. Pick a 2. ____________

9. Pick an even number. ____________

2 4 8 6 6 8 4 2

Draw a spinner for each probability.

10. Likely but not certain to land on a 5

11. Unlikely but not impossible to land on red

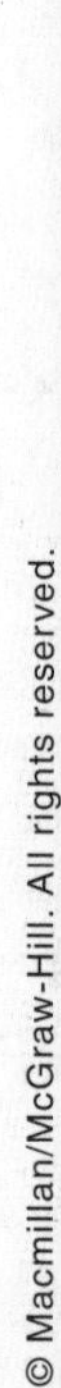

Name ______________________________

Explore Finding Outcomes

List the possible outcomes for each. Then do the experiment.
Record the outcomes in a line plot and a bar graph.

1. Place 3 red cubes, 2 blue cubes, and 1 yellow cube in a paper bag. Take a cube from the bag, record it, and replace it in the bag. Repeat 20 times.

Possible outcomes: ______________________________

Draw your line plot here.

Draw your bar graph on the grid paper.

2. Place 1 dime, 1 nickel, and 1 penny in a paper bag. Take a coin from the bag, record it, and replace it in the bag. Repeat 20 times.

Possible outcomes: ______________________________

Draw your line plot here.

Draw your bar graph on the grid paper.

Name ____________________

Problem Solving: Strategy
Make an Organized List

P 26-7 PRACTICE

Use an organized list to solve.

1. Diane is buying a bag. She can choose a large, medium, or small bag. The bag comes in leather or canvas. The bag comes with or without a strap. How many different bags are there?

____

2. Sandy wants to buy breakfast. She has a choice of a blueberry, a corn, or a bran muffin to eat. She has a choice of apple, orange, or grapefruit juice to drink. How many different breakfasts are possible?

____

3. Mr. Bevin is flying to Tokyo. The airline offers flights at 8:00 A.M. and 6:00 P.M. On each flight, there are first class, business class, and coach tickets. How many different choices are there for Mr. Bevin?

____

4. Dan can buy a vest in blue, black, green, or white. He can choose a V-neck or a crew neck. He can also choose a vest with or without a pocket. How many different kinds of vests are there?

____

Mixed Strategy Review

5. Tara gets to her hotel at 11:15 A.M. She was traveling for 1 hour 45 minutes. At what time did Tara start traveling?

____

6. Jamal spends $379 on plane fare. He spends $450 on a hotel room and other trip expenses. How much money does Jamal spend?

____

7. **Art** Joe wants to make a picture. He can use oil paints, watercolor, or ink. He can make a large or a small picture. How many different choices does Joe have?

____

8. **Write a problem** which can be solved by making an organized list. Share it with others.

____

____

____

Name ___________________________

Explore Fractions and Decimals

P 27-1 PRACTICE

Shade as indicated.
Then write a fraction and a decimal for each shaded part.

1. Shade 4 parts. **2.** Shade 8 parts. **3.** Shade 1 part. **4.** Shade 6 parts.

_____ = _____ _____ = _____ _____ = _____ _____ = _____

5. Shade 15 parts. **6.** Shade 35 parts. **7.** Shade 5 parts. **8.** Shade 55 parts.

_____ = _____ _____ = _____ _____ = _____ _____ = _____

Write a fraction and a decimal for each shaded part.

9. _____ **10.** _____ **11.** _____ **12.** _____

13. _____ **14.** _____ **15.** _____ **16.** _____

Name ____

Fractions and Decimals

P 27-2 PRACTICE

Write a decimal for each.

1. 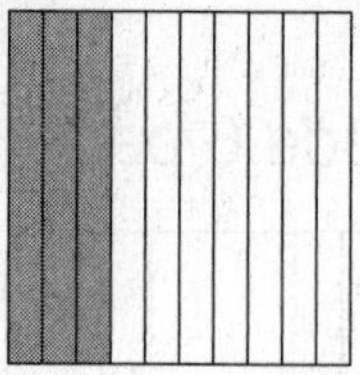 ________ **2.** 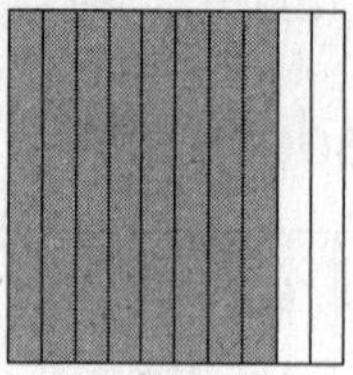 ________ **3.** 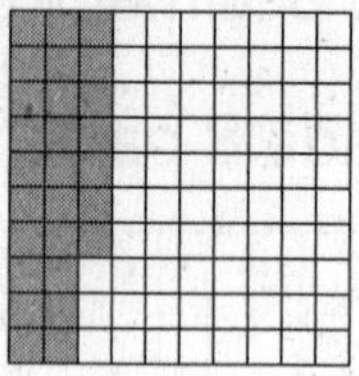 ________ **4.** ________

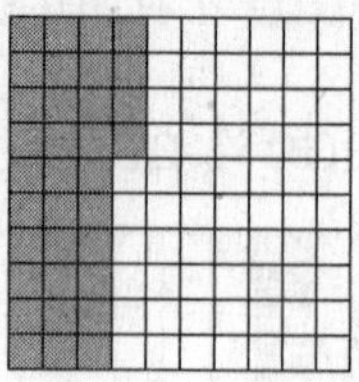

5. $\frac{5}{10}$ = ________ **6.** $\frac{1}{10}$ = ________ **7.** $\frac{9}{10}$ = ________ **8.** $\frac{4}{10}$ = ________

9. $\frac{25}{100}$ = ________ **10.** $\frac{16}{100}$ = ________ **11.** $\frac{9}{100}$ = ________ **12.** $\frac{7}{100}$ = ________

13. six tenths ________

14. eight tenths ________

15. twenty-three hundredths ________

16. nine hundredths ________

17. 45¢ ________

18. 92¢ ________

19. 2¢ ________

20. 57¢ ________

21. 16¢ ________

22. 39¢ ________

Problem Solving

Solve.

23. There are 10 children at the Sunnyside Preschool. Seven children are younger than 4 years old. Write a fraction and a decimal for the number of children who are younger than 4 years old.

____

24. There are 100 third grade students at Otsego Elementary School. Forty-nine of the students are girls. Write a fraction and a decimal for the number of girls.

____

Name ______________________________

Decimals Greater Than One

P 27-3 PRACTICE

Write each decimal.

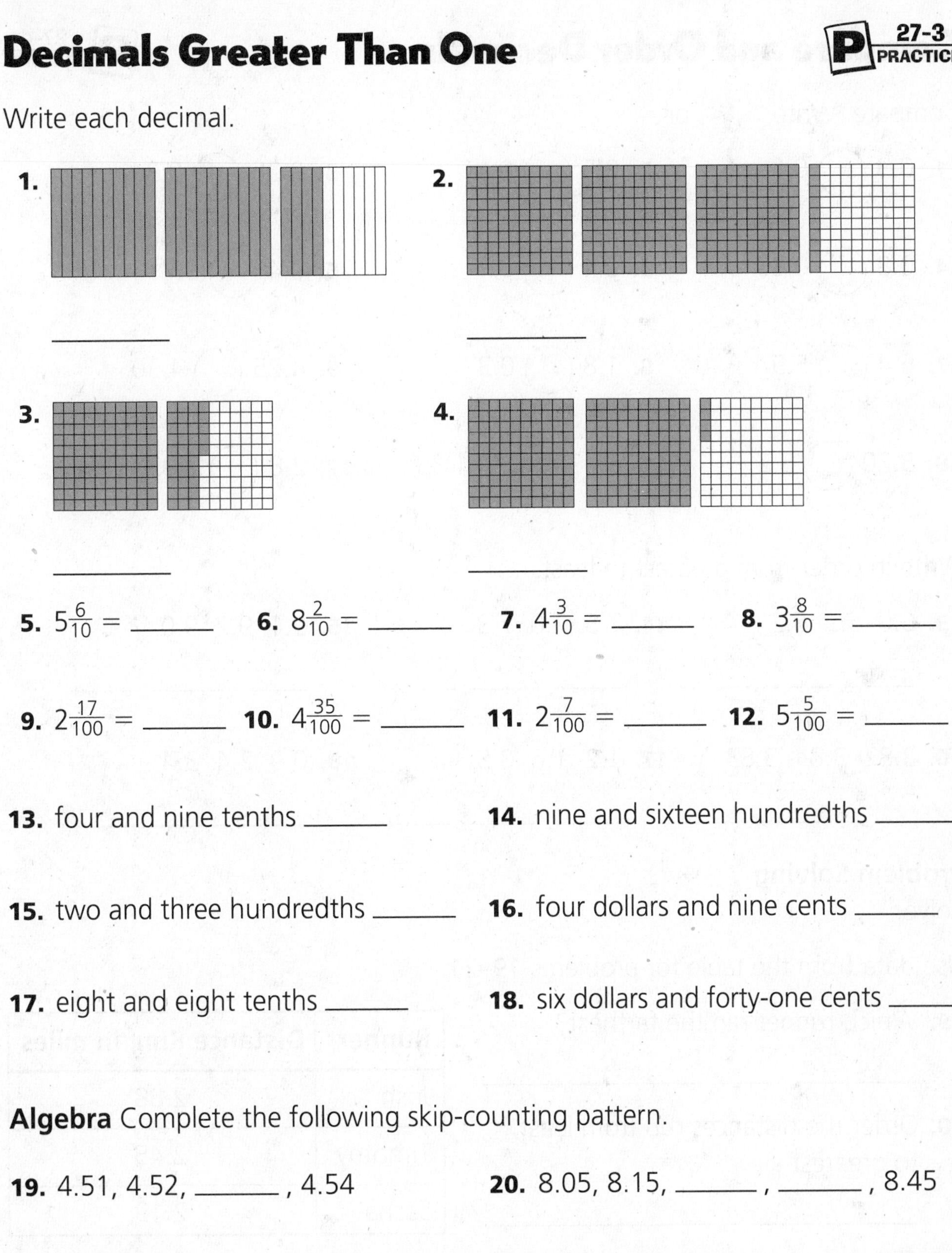

1. ________

2. ________

3. ________

4. ________

5. $5\frac{6}{10}$ = ______ **6.** $8\frac{2}{10}$ = ______ **7.** $4\frac{3}{10}$ = ______ **8.** $3\frac{8}{10}$ = ______

9. $2\frac{17}{100}$ = ______ **10.** $4\frac{35}{100}$ = ______ **11.** $2\frac{7}{100}$ = ______ **12.** $5\frac{5}{100}$ = ______

13. four and nine tenths ______

14. nine and sixteen hundredths ______

15. two and three hundredths ______

16. four dollars and nine cents ______

17. eight and eight tenths ______

18. six dollars and forty-one cents ______

Algebra Complete the following skip-counting pattern.

19. 4.51, 4.52, ______, 4.54

20. 8.05, 8.15, ______, ______, 8.45

21. 6.21, 6.31, ______, ______, 6.61

22. 1.35, 1.36, ______, ______, 1.39

Name ____

Compare and Order Decimals

Compare. Write >, <, or =.

1. 0.3 ◯ 0.6 **2.** 0.5 ◯ 0.1 **3.** 0.25 ◯ 0.34

4. 0.8 ◯ 0.80 **5.** 0.4 ◯ 0.1 **6.** 0.4 ◯ 0.7

7. 6.4 ◯ 5.9 **8.** 1.8 ◯ 0.3 **9.** 4.25 ◯ 4.30

10. 5.20 ◯ 5.2 **11.** $1.34 ◯ $1.35 **12.** 2.8 ◯ 2.83

Write in order from greatest to least.

13. 6.0 5.9 6.2 **14.** 7.5 7.0 7.3 **15.** 9.1 9.2 9.0

_____, _____, _____ _____, _____, _____ _____, _____, _____

16. 3.82 3.84 3.83 **17.** 1.2 1.6 0.5 **18.** 0.9 2.4 3.4

_____, _____, _____ _____, _____, _____ _____, _____, _____

Problem Solving

Solve.

Use data from the table for problems 19–21.

19. Which runner ran the farthest?

____

20. Order the distances run from least to greatest.

____

21. Kelly ran 2.8 miles. Did she run more or less than Josh? How do you know?

____

Runner	Distance Run, in miles
Josh	2.08
Timothy	2.45
Sasha	2.19

Name ________________________________

Problem Solving: Skill

Choose an Operation

Solve.

1. Jacob spends \$96 for 4 wheels. What is the cost for each wheel?

____

2. Jon spends \$12.50 on a book about inventions and \$6.75 on a poster. How much does he spend in all?

____

3. Hiroshi buys electrical parts for \$32.48 and other parts for \$19.99. How much money does he spend?

____

4. Sid sells his invention for \$49.25. Liza sells her invention for \$15.60 more than that. How much does Liza receive for her invention?

____

5. Lyn spends \$16.30 on wood and \$3.99 on wire. How much does she spend?

____

6. Teddy buys 3 tickets to the Inventors' Meeting. Tickets cost \$8.50 each. How much does Teddy spend?

____

Mixed Strategy Review

7. A book on Thomas Edison costs \$19.49. A book on Albert Einstein costs \$12.95. What is the difference between the prices of the two books?

____

8. Fiona buys two books for \$4.99 each and another for \$12.50. Can she pay with a twenty-dollar bill? Explain your answer.

____

____

Use with Grade 3, Chapter 27, Lesson 5, pages 618–619.

Name ____________________

Explore Adding Decimals

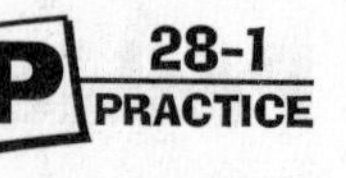

Add.

1. 0.3 + 0.8 = _____

2. 1.3 + 1.0 = _____

3. 0.25 + 0.47 = _____

4. 1.19 + 0.61 = _____

Add. Shade models to show your addition.

5. 0.3 + 0.7 = _____

6. 0.8 + 0.7 = _____

7. 0.25 + 0.25 = _____

8. 1.16 + 1.34 = _____

Add. Use models if you wish.

9. 0.8 + 0.8

10. 1.3 + 0.7

11. 1.32 + 2.75

12. 0.8 + 0.4

13. 1.24 + 2.05

14. 2.16 + 1.72

Name ______________________________

Add Decimals

Add.

1. 0.4 + 0.9	**2.** 0.5 + 0.6	**3.** 1.3 + 0.2	**4.** 3.5 + 2.8	**5.** 4.7 + 2.6
6. 0.24 + 0.43	**7.** 0.58 + 0.29	**8.** 2.64 + 0.39	**9.** $2.46 + 1.93	**10.** $7.45 + 2.16
11. 2.45 + 1.60	**12.** 2.2 + 1.8	**13.** $3.29 + 1.48	**14.** 6.28 + 0.17	**15.** 5.32 + 1.78

16. 0.7 + 0.7 = ______

17. 0.9 + 1.3 = ______

18. 1.63 + 2.41 = ______

19. 1.55 + 1.84 = ______

20. $4.37 + $1.99 = ______

21. 7.29 + 0.16 = ______

Problem Solving

Solve.

22. Dave saved $0.75. He finds $2.35 in his pocket. How much does he have now?

____

23. Dave spends $2.49 for a card. He spends another $0.58 for postage. How much does he spend in all?

____

24. Kari spent $3.25 for lunch. Melissa spent $4.95. How much money did they spend altogether?

____

25. Alan and Li each spent $4.45 for lunch. How much did they spend altogether?

____

Name ____________________

Problem Solving: Strategy
Solve a Simpler Problem

Solve.

1. Tickets to the Science Center cost \$6.95 for adults and \$3.95 for children. How much does a family of 2 adults and 4 children pay for tickets?

   ____

2. The Yuen family stops in the gift shop. Science Center pens cost \$3.65. Science Center buttons cost \$1.95. How much does it cost to buy 2 pens and 3 buttons?

   ____

3. Workers at the Science Center rope off a rectangular space. The space has sides of 5.8 meters and 8.9 meters. How much rope do they need?

   ____

4. Lana's home is 0.9 miles away from the bus stop. The ride from the bus stop to the Science Center is 5.6 miles. Lana walks to the bus stop and takes the bus to the Science Center. She returns home the same way. How many miles does she travel in all?

   ____

Mixed Strategy Review

5. Nell, Barry, Chet, and Jill are in line for a movie on Alexander Graham Bell. The first person in line is a boy. Barry is ahead of Nell, but not ahead of Jill. List the names in order from first to last in line.

   ____

6. **Write a problem** that you could use a simpler problem to solve. Share it with others.

   ____

   ____

   ____

Name ____________________

Explore Subtracting Decimals

Subtract.

1. 0.7 − 0.2 = ______

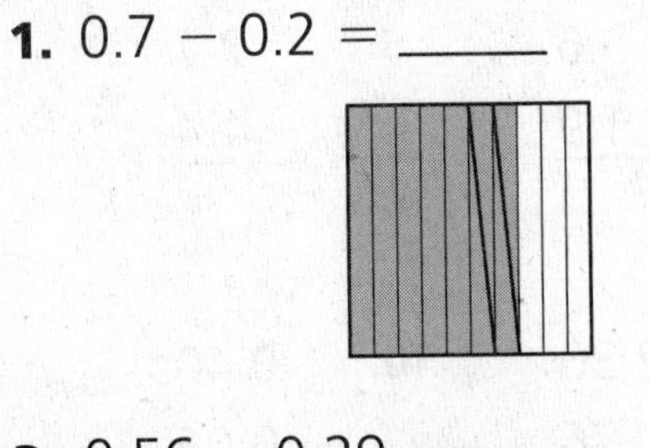

2. 1.4 − 0.8 = ______

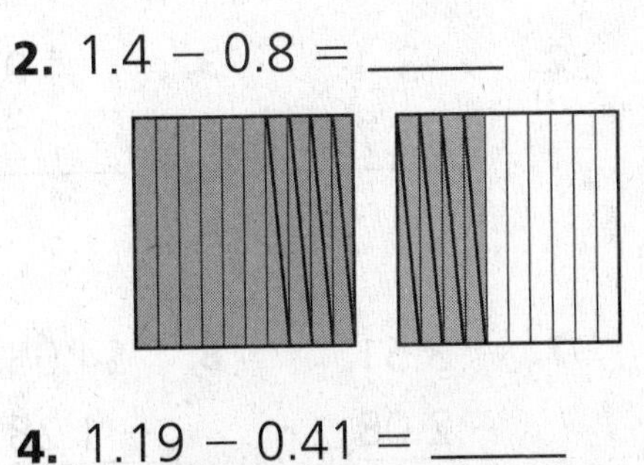

3. 0.56 − 0.29 = ______

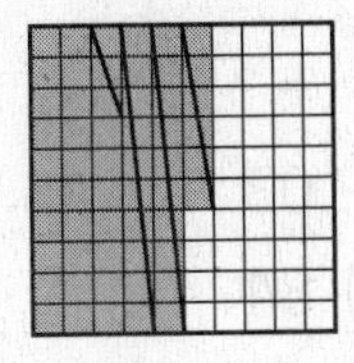

4. 1.19 − 0.41 = ______

Subtract. Show your subtraction on the models.

5. 0.8 − 0.1 = ______

6. 1.2 − 0.5 = ______

7. 0.62 − 0.19 = ______

8. 1.84 − 0.62 = ______

Subtract. Use models if you wish.

9. 1.6 − 1.4

10. 1.25 − 0.75

11. 1.5 − 0.8

12. 1.75 − 1.50

Name ____________________

Subtract Decimals

Subtract.

1. 2.3 − 0.7	**2.** 3.4 − 1.7	**3.** 4.0 − 1.2	**4.** 5.6 − 1.9	**5.** 2.7 − 1.4
6. 2.46 − 1.28	**7.** 3.51 − 2.08	**8.** $4.05 − 1.28	**9.** 2.30 − 0.76	**10.** 0.40 − 0.06
11. 0.30 − 0.12	**12.** 2.41 − 1.56	**13.** $1.25 − 0.50	**14.** 2.03 − 1.84	**15.** 4.70 − 1.29
16. 6.3 − 1.4	**17.** 2.4 − 1.8	**18.** 4.36 − 2.18	**19.** $1.04 − 0.65	**20.** 2.40 − 1.38

Problem Solving

Solve.

21. One model robot takes 1.38 seconds to do a job. A newer model is 0.7 second faster. How long does it take the new model to do the job?

____

22. A ticket for the robot show at the science museum costs $5. John has saved $3.84 already. How much more does John need for his ticket?

____

23. Gavin spent $5.85 for a souvenir. Marta spent $8.10 for her souvenir. How much more did Marta spend than Gavin?

____

24. Drew bought a Speed Scooter game for $9.00. He bought a RoboMaze game for $7.85. How much more did Speed Scooter cost than RoboMaze?

____

Summer Skills Refresher

Summer Skills

Facts from the Geography Files

When you look at a map of Florida, you can see that it lies between the Atlantic Ocean and the Gulf of Mexico.

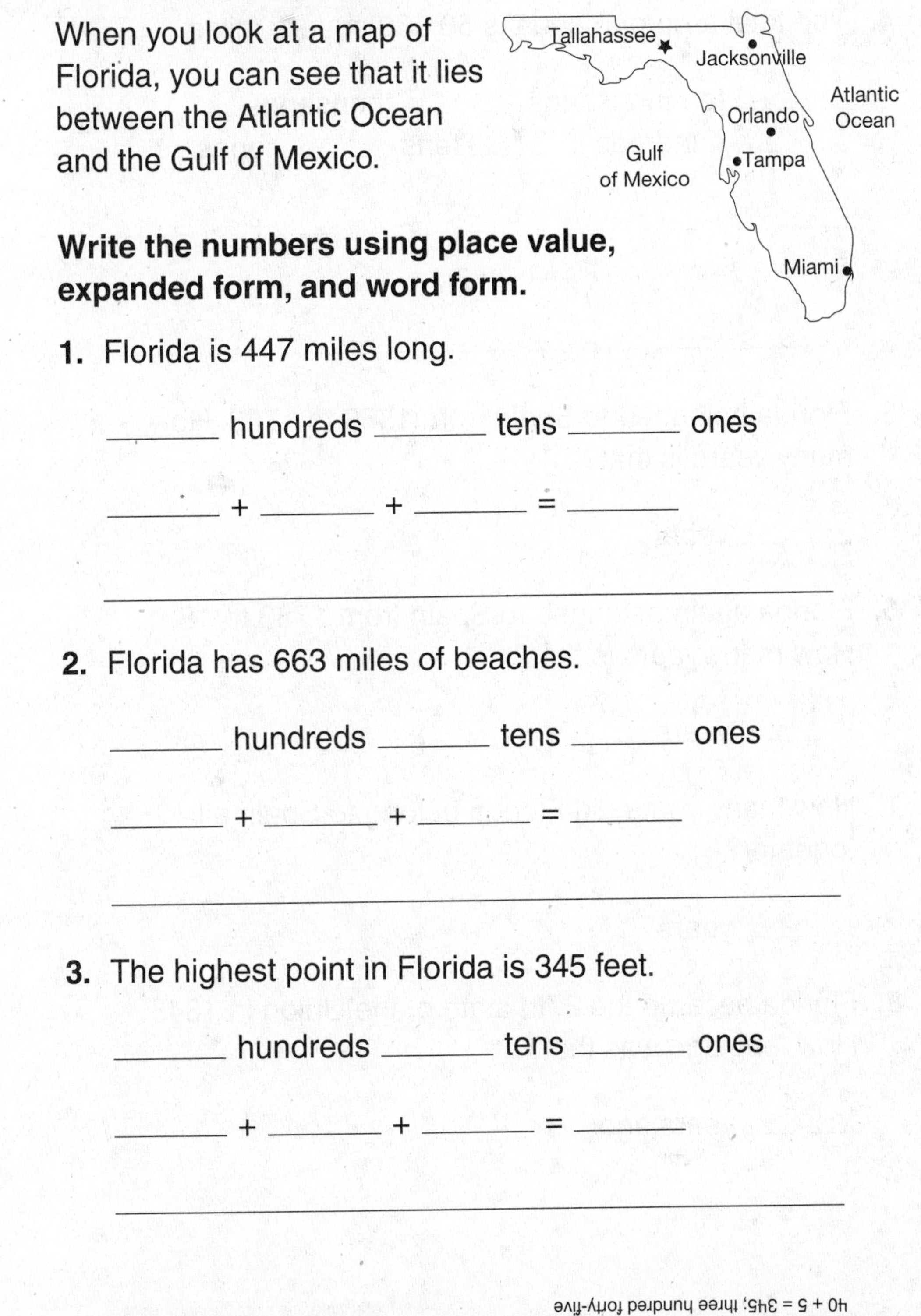

Write the numbers using place value, expanded form, and word form.

1. Florida is 447 miles long.

   ______ hundreds ______ tens ______ ones

   ______ + ______ + ______ = ______

   ____

2. Florida has 663 miles of beaches.

   ______ hundreds ______ tens ______ ones

   ______ + ______ + ______ = ______

   ____

3. The highest point in Florida is 345 feet.

   ______ hundreds ______ tens ______ ones

   ______ + ______ + ______ = ______

   ____

Answers: 1. 4 hundreds 4 tens 7 ones; 400 + 40 + 7 = 447; four hundred forty-seven; 2. 6 hundreds 6 tens 3 ones; 600 + 60 + 3 = 663; six hundred sixty-three; 3. 3 hundreds 4 tens 5 ones; 300 + 40 + 5 = 345; three hundred forty-five

4. The total area of Florida is 58,560 square miles.

_______ ten thousands _______ thousands
_______ hundreds _______ tens _______ ones

_______ + _______ + _______ + _______ + _______
= _______

____

5. Florida belonged to Spain from 1565 to 1763. How many years is that?

_______ years

6. Florida again belonged to Spain from 1783 to 1821. How many years is that?

_______ years

7. How many years did Florida belong to Spain all together?

_______ years

8. Florida became the 27th state of the Union in 1845. How long ago was that?

_______ years ago

Answers: **4.** 5 ten thousands 8 thousands 5 hundreds 6 tens 0 ones; 50,000 + 8,000 + 500 + 60 = 58,560; fifty-eight thousand five hundred sixty; **5.** 198 years; **6.** 38 years; **7.** 236 years; **8.** In 2004, it was 159 years ago.

Summer Skills

Beneath the Water

Around the southeastern coast of the United States, there are sharks, dolphins, alligators, and manatees.

Circle the more reasonable answer.

1. The blubber layer on a dolphin's body is this thick.

   1 feet   1 mile   1 inch

2. There are about 40 different kinds of dolphins. Dolphins can grow this long.

   13 feet   13 inches   13 miles

3. The orca or killer whale can grow to be this long.

   2 inches   324 inches   4 inches

4. The American alligator can be this long.

   4 millimeters   4 centimeters   4 meters

Answers: 1. inch; 2. 13 feet; 3. 324 inches; 4. 4 meters

Circle the more reasonable answer.

5. A manatee is a mammal that lives in warm waters. The manatee can be this long.

   300 meters   300 centimeters   300 millimeters

6. The bull shark can weigh this much.

   285 tons   285 pounds   285 ounces

7. The dolphin can weigh this much.

   450 pounds   450 ounces   450 grams

8. The American alligator can weigh this much.

   181 grams   181 kilograms   181 tons

9. The manatee can weigh this much.

   1,360 kilograms
   1,360 grams
   1,360 milligrams

10. The American alligator can weigh this much.

   400 ounces   400 pounds   400 tons

Answers: **5.** 300 centimeters; **6.** 285 pounds; **7.** 450 pounds; **8.** 181 kilograms; **9.** 1,360 kilograms; **10.** 400 pounds

Summer Skills

Flipping the Alligator

The American alligator is interesting to observe, but it is best not to get too close. The alligator moves quickly for its size. This alligator is on the move. Use the words in the box to describe its moves.

slide
turn
flip

1. This alligator is sneaking away. What move is it making here?

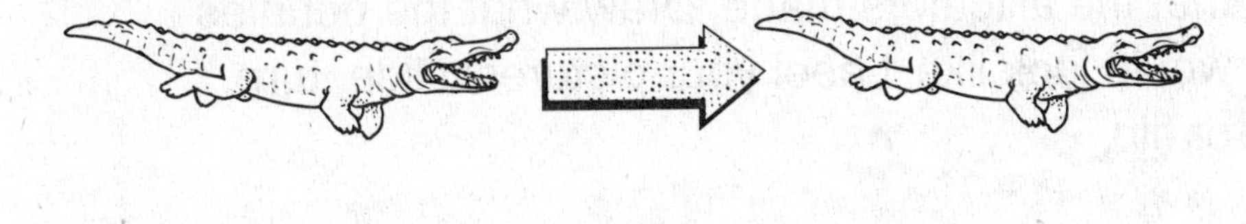

____

2. This alligator is curious. What move is this alligator making?

____

Answers: 1. slide; 2. turn

3. What move is this alligator making?

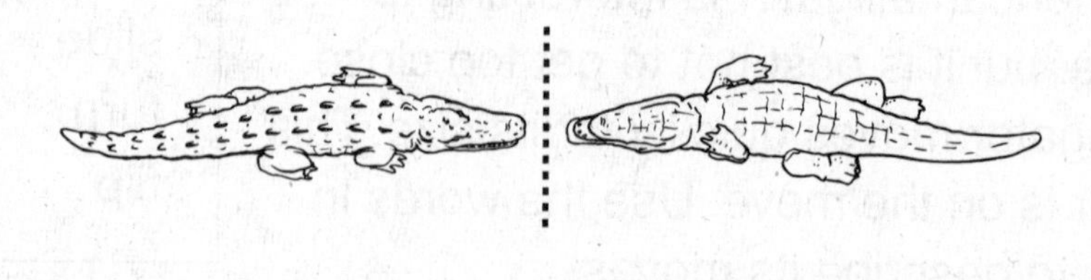

---

4. Your turn! Use a penny and show the same moves that the alligators made. Draw what the pennies would look like. Label your pictures: slide, turn, or flip.

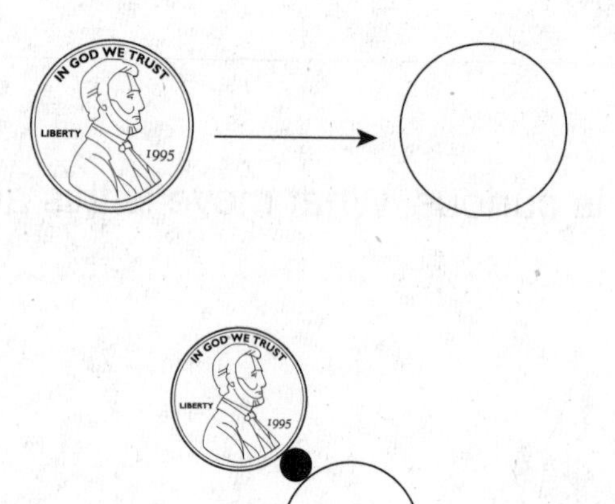

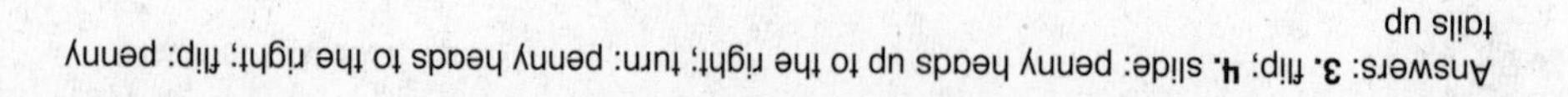

Summer Skills

How Much Does It Cost?

Maija and her parents are going to visit her grandparents in California.

1. Maija, her parents, and her grandparents went to a butterfly park. Maija knows that each adult ticket cost $10.95. Maija's dad got 20¢ back from $50. How much did Maija's ticket cost?

   $10.95 + $10.95 + $10.95 + $10.95 + $_______
   = $50 – $0.20

2. Maija's family went to a farmer's market. Maija bought a smoothie, and her mom bought lemonade for $2.45. The two items cost $6.40. How much did the smoothie cost?

   $_______ + $2.45 = $6.40

3. At the Pacific Castle, Maija's grandfather paid for the tickets with $35. He got $1.30 back. How much did the tickets cost all together?

   $1.30 + $_______ = $35.00

4. Maija's mom and dad paid $7.75 each. How much did their tickets cost in all?

   $_______ + $_______ = $_______

Answers: 1. $6; 2. $3.95; 3. $33.70; 4. $15.50

5. Maija's grandparents paid $1.15 less than Maija's parents for each of their senior citizen tickets. How much did each ticket cost?

$______ + $1.15 = $7.75

6. How much did it cost for Maija to get into the Pacific Castle?

$______ + $______ + $7.75 + $7.75 + $______
= $35.00 – $1.30

7. The last day they all went to visit Palm Springs. A tour bus cost $6.00 for a child's ticket. The price for all tickets was $56.00. How much does an adult ticket cost?

____

Answers: 5. $6.60; 6. 5.00; $6.60 + $6.60 + $7.75 + $7.75 + $5.00 = $35 – $1.30; 7. $12.50; $56.00 – $6.00 = $50, $50.00 ÷ 4 = 12.50

Summer Skills

Pelicans and Ibises

Pelicans and Ibises are unusual and interesting birds. There are several varieties of each.

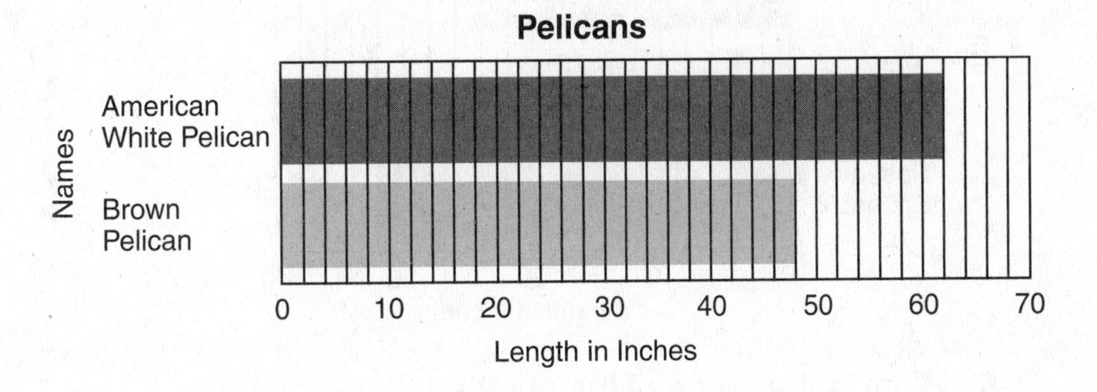

1. The length of a bird is measured from the tip of the bill to the end of the tail feathers. How long is the American white pelican?

   _______ inches

2. How long is the brown pelican?

   _______ inches

3. How much longer is the white pelican than the brown pelican?

   _______ inches

4. About how many feet is that?

   _______ feet

Answers: 1. 62 inches; 2. 48 inches; 3. 14 inches; 4. a little more than 1 foot

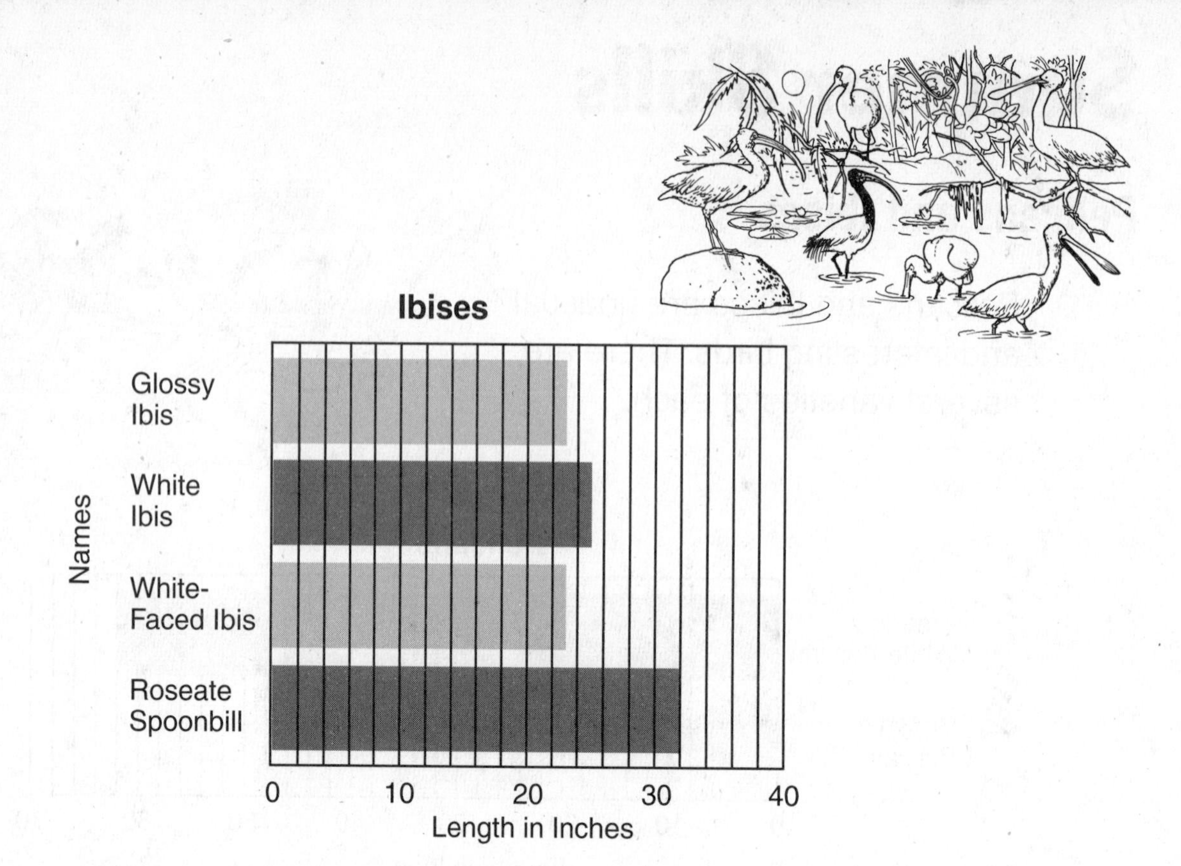

5. What is the range of the numbers?

____

6. What is the mode?

____

7. What is the median?

____

8. If you compare the pelicans and the ibises, which kind of bird is longer? How much longer?

____

____

Answers: **5.** 9; **6.** 23; **7.** the number between 23 and 25, 24; **8.** The American white pelican, 30 inches